QUARKS, GLUONS AND LATTICES

MICHAEL CREUTZ

Brookhaven National Laboratory

CAMBRIDGE UNIVERSITY PRESS
Cambridge, New York, Melbourne, Madrid, Cape Town, Singapore,
São Paulo, Delhi, Dubai, Tokyo, Mexico City

Cambridge University Press
The Edinburgh Building, Cambridge CB2 8RU, UK

Published in the United States of America by
Cambridge University Press, New York

www.cambridge.org
Information on this title: www.cambridge.org/9780521315357

© Cambridge University Press 1983

First published 1983
First paperback edition (with corrections) 1985
Reprinted 1986,1988, 1990, 1997

A catalogue record for this publication is available from the British Library

Library of Congress Cataloguing in Publication Data

ISBN 978-0-521-24405-6 Hardback
ISBN 978-0-521-31535-7 Paperback

Contents

Preface

The last few years have seen a dramatic upsurgence in interdisciplinary activity between solid state and particle physics. This arose primarily through the use of lattice cutoffs to study non-perturbative phenomena in the gauge theory of the strong interactions. However, the standard textbook treatments of field theory rely on more traditional perturbative techniques. This book is an attempt to introduce lattice techniques to the particle physicist with a basic background in relativistic quantum mechanics. This work is not intended to be a review of the latest developments, which are rapidly evolving, but rather an exposition of some of the more established methods.

The presentation is in the framework of particle physics. Solid state physicists may be interested in why high energy theorists are borrowing many of their ideas, but they should not expect this book to address subjects primarily of interest to their field. Thus important topics such as two-dimensional spin models, critical exponents, and fixed point phenomenology are only superficially mentioned.

I am grateful to the International School for Advanced Study in Trieste, Italy, and to my colleagues at Brookhaven for the opportunity to present series of lectures based on portions of this book.

1
Quarks and gluons

Our prime candidate for a fundamental theory of strong hadronic forces is a model of quarks interacting through the exchange of non-Abelian gauge fields. The quark model represents a new level of substructure within hadronic particles such as the proton. We have several compelling reasons to believe in this next layer of matter.

First, the large cross sections observed in deeply inelastic lepton–hadron scattering indicate important structure at distance scales of less than 10^{-16} centimeters, whereas the overall proton electromagnetic radius is of order 10^{-13} centimeters. The angular dependences observed in these experiments suggest that the underlying charged constituents carry half-integer spin. These studies have raised the question of whether it is theoretically possible to have pointlike objects in a strongly interacting theory. Asymptotically free non-Abelian gauge interactions offer this hope (Perkins, 1977).

A second impetus for a theory of quarks lies in low energy hadronic spectroscopy. Indeed, it was the successes of the eightfold way (Gell-Mann and Ne'eman, 1964) which originally motivated the quark model. We now believe that the existence of two 'flavors' of low mass quarks lies at the heart of the isospin symmetry in nuclear physics. Adding a somewhat heavier 'strange' quark to the theory gives rise to the celebrated multiplet structure in terms of representations of the group $SU(3)$.

Third, we have further evidence for compositeness in the excitations of the low-lying hadrons. Particles differing in angular momentum fall neatly into place on the famous 'Regge trajectories' (Collins and Squires, 1968). In this way families of states group together as orbital excitations of some underlying system. The sustained rising of these trajectories with increasing angular momentum points toward strong long-range forces. This originally motivated the stringlike models of hadrons.

Finally, the idea of quarks became incontrovertible with the discovery of the 'hydrogen atoms' of elementary particle physics. The intricate spectroscopy of the charmonium and upsilon families is admirably explained in potential models for non-relativistic bound states of heavy quarks (Eichten *et al.*, 1980).

1

Despite these successes of the quark model, an isolated quark has never been observed. (Some hints of fractionally charged macroscopic pieces of matter may eventually prove to contain unbound quarks, or might be a sign of some new and even more exciting type of matter (LaRue, Phillips and Fairbank, 1981).) These basic constituents of matter do not copiously appear as free particles emerging from present laboratory experiments. This is in marked contrast to the empirical observation in hadronic physics that anything which can be created will be. The difficulty in producing quarks has led to the speculation of an exact confinement. Indeed, it may be simpler to imagine a constituent which can never be produced than an approximate imprisonment relying on some unnaturally effective suppression factor in a theory seemingly devoid of any large dimensionless parameters.

But how can we ascribe any reality to an object which cannot be produced? Are we just dealing with some sort of mathematical trick? We will now argue that gauge theories potentially possess a simple mechanism for giving constituents infinite energy when in isolation. In this picture a quark–antiquark pair will experience an attractive force which remains non-vanishing even for asymptotically large separations. This linearly rising long-distance potential energy forms the basis of essentially all models of quark confinement.

We begin by coupling the quarks to a conserved 'gluo-electric' flux. In usual electromagnetism the electric field lines thus produced spread and give rise to the inverse square law Coulombic field. If in our theory we can now somehow eliminate massless fields, then a Coulombic spreading will no longer be a solution to the equations. If in removing the massless fields we do not destroy the Gauss law constraint that the quarks are the sources of electric flux, the electric lines must form into tubes of conserved flux, schematically illustrated in figure 1.1. These tubes will only end on the quarks and their antiparticles. A flux tube is a real physical object carrying a finite energy per unit length. This is the storage medium for the linearly rising interquark potential (Kogut and Susskind, 1974).

A simple model for this phenomenon is a type II superconductor containing magnetic monopole impurities. Because of the Meissner effect (Meissner and Ochsenfeld, 1933), a superconductor does not admit magnetic fields. However, if we force a hypothetical magnetic monopole into the system, its lines of magnetic flux must go somewhere. Here the role of the 'gluo-electric' flux is played by the magnetic field, which will bore a tube of normal material through the superconductor until it ends on an antimonopole or it leaves the boundary of the system. Such flux tubes have been experimentally observed in applied magnetic fields (Huebner and Clem, 1974).

Another example of this mechanism occurs in the bag model (Chodos *et al.*, 1975). Here the gluonic fields are unrestricted in the baglike interior of a hadron but forbidden by ad hoc boundary conditions from extending outside. In attempting to extract a single quark from a proton, one would draw out a long skinny bag carrying the gluo-electric flux of the quark back to the remaining constituents.

Fig. 1.1. A flux tube from a quark to an antiquark.

The above models may be interesting phenomenologically, but they are too arbitrary to be considered as the basis for fundamental theories. In their search for a more elegant model, theorists have been drawn to non-Abelian gauge fields. This dynamical system of coupled gluons begins like electrodynamics with a set of massless gauge fields interacting with the quarks. Using the freedom of an internal symmetry, the action includes self-couplings of the gluons. The bare massless fields are all charged with respect to each other. The confinement conjecture is that this input theory of massless charged particles is unstable to a condensation of the vacuum to a state in which only massive excitations can propagate. In such a state the gluonic flux around quarks should form into the tubes needed for linear confinement. Much of the recent effort in elementary particle theory has gone into attempts to show that this indeed takes place.

The confinement phenomenon makes the theory of the strong interactions qualitatively different from theories of the electromagnetic and weak forces. The fundamental fields of the Lagrangian do not manifest themselves in free hadronic spectrum. In not observing free quarks and gluons, we are led to the conjecture that all observable strongly interacting particles are gauge singlet bound states of these fundamental constituents.

In the usual quark model baryons are bound states of three quarks. Thus the gauge group should permit singlets to be formed from three objects in the fundamental representation. This motivates the use of $SU(3)$ as the underlying group of the strong interactions. This internal symmetry must not be confused with the broken $SU(3)$ represented in spectroscopic multiplets. Ironically, one of the original motivations for quarks has now become an accidental symmetry. The symmetry considered here is hidden behind the confinement mechanism, which only permits us to observe singlet states.

For the presentation in this book we assume, perhaps too naively, that the nuclear interactions can be considered in isolation from the much weaker effects of electromagnetism, weak interactions, and gravitation. This does not preclude the possible application of the techniques presented here to the other interactions. Indeed, grand unification may be crucial for a consistent theory of the world. To describe physics at normal laboratory energies, however, only for the strong interaction must we go beyond well-established perturbative methods. Thus we frame our discussion around quarks and gluons.

2
Lattices

The best evidence we have for confinement in a non-Abelian gauge theory of the strong interactions comes by way of Wilson's (1974) formulation on a space-time lattice. At first this prescription seems a little peculiar because the vacuum is not a crystal. Indeed, experimentalists work daily with relativistic particles showing no deviations from the continuous symmetries of the Lorentz group. Why, then, have theorists in recent years spent so much time describing field theory on the scaffolding of a space-time lattice?

The lattice represents a mathematical trick. It provides a cutoff removing the ultraviolet infinities so rampant in quantum field theory. As with any regulator, it must be removed after renormalization. Physics can only be extracted in the continuum limit, where the lattice spacing is taken to zero.

But infinities and the resulting need for renormalization have been with us since the beginnings of relativistic quantum mechanics. The program for electrodynamics has had immense success without recourse to discrete space. Why reject the time-honored perturbative renormalization procedures in favor of a new cutoff scheme?

We are driven to the lattice by the rather unique feature of confinement in the strong interactions. This phenomenon is inherently non-perturbative. The free theory with vanishing coupling constant has no resemblance to the observed physical world. Renormalization group arguments, to be presented in detail in later chapters, indicate severe essential singularities when hadronic properties are regarded as functions of the gauge coupling. This contrasts sharply with the great successes of quantum electrodynamics, where perturbation theory was central. Most conventional regularization schemes are based on the Feynman expansion; some process is calculated until a divergence is met in a particular diagram, and this divergence is then removed. To go beyond the diagrammatic approach, one needs a non-perturbative cutoff. Herein lies the main virtue of the lattice, which directly eliminates all wavelengths less than twice the lattice spacing. This occurs before any expansions or approximations are begun.

On a lattice, a field theory becomes mathematically well-defined and can

5

be studied in various ways. Lattice perturbation theory, although somewhat awkward, recovers all the conventional results of other regularization schemes. Discrete space-time, however, is particularly well-suited for a strong coupling expansion. Remarkably, confinement is automatic in this limit where the theory reduces to one of quarks on the ends of strings with a finite energy per unit length. Most recent research has concentrated on showing that this phenomenon survives the continuum limit.

A lattice formulation emphasizes the close connections between field theory and statistical mechanics. Indeed, the strong coupling treatment is equivalent to a high temperature expansion. The deep ties between these disciplines are manifest in the Feynman path integral formulation of quantum mechanics (Feynman, 1948; Dirac, 1933, 1945). In Euclidian space, a path integral is equivalent to a partition function for an analogous statistical system. The square of the field theoretical coupling constant corresponds directly to the temperature. Thus, the particle physicist has available the full technology of the condensed matter theorist.

Confinement is natural in the strong coupling limit of the lattice theory; however, this is not the region of direct physical interest, for which a continuum limit is necessary. The coupling constant on the lattice represents a bare coupling at a length scale of the lattice spacing. Non-Abelian gauge theories possess the property of asymptotic freedom, which means that in the short distance limit the effective coupling goes to zero. This remarkable phenomenon allows predictions for the observed scaling behavior in deeply inelastic collisions. Indeed, this was one of the original motivations for a non-Abelian gauge theory of the strong interactions. The consequence for the lattice theory, however, is that the bare coupling must be taken to zero as the lattice spacing decreases towards the continuum limit. Thus we are inevitably led out of the high temperature regime and into a low temperature domain. Along the way in a general statistical system one might expect to encounter phase transitions. Such qualitative shifts in the physical characteristics of a system can only hamper the task of showing confinement in the non-Abelian theory. In later chapters we will present evidence that such troublesome transitions can be avoided in the four-dimensional $SU(3)$ gauge theory of the nuclear force.

Although our ultimate goal with lattice gauge theory is an understanding of hadronic physics, many interesting phenomena arise which are peculiar to the lattice. We will see non-trivial phase structure occurring in a variety of models, some of which do not correspond to any continuum field theory. The lattice formulation is highly non-unique and thereby spurious transitions can be alternately introduced and removed. We will also see that the statistical mechanics of gauge models displays curious analogies with magnetic systems in half the number of space-time dimensions. Even quantum electrodynamics shows interesting structure in certain lattice formulations. This rich spectrum of phenomena has led to the recent popularity of lattice field theories and motivates this book.

3

Path integrals
and statistical mechanics

The Feynman path integral formulation of quantum mechanics reveals deep connections with statistical mechanics. This chapter is concerned with this relationship for the simple case of a non-relativistic particle in a potential. Starting with a partition function representing a path integral on an imaginary time lattice, we will show how a transfer matrix formalism reduces the problem to the diagonalization of an operator in the usual quantum mechanical Hilbert space of square integrable functions (Creutz, 1977). In the continuum limit of the time lattice, we obtain the canonical Hamiltonian. Except for our use of imaginary time, this treatment is identical to that in Feynman's early work (Feynman, 1948).

We begin with the Lagrangian for a free particle of mass m moving in potential $V(x)$

$$L(x, \dot{x}) = K(\dot{x}) + V(x), \tag{3.1}$$

$$K(\dot{x}) = \tfrac{1}{2} m \dot{x}^2, \tag{3.2}$$

where $\dot{x}$ is the time derivative of the coordinate x. Velocity-dependent potentials are beyond the scope of this book. Note the unconventional relative positive sign between the two terms in eq. (3.1). This is because we formulate the path integral directly in imaginary time. This improves mathematical convergence, yet leaves us with the usual Hamiltonian for diagonalization.

For any trajectory we have an action

$$S = \int \mathrm{d}t \, L(\dot{x}(t), x(t)), \tag{3.3}$$

which appears in the path integral

$$Z = \int [\mathrm{d}x(t)] \, \mathrm{e}^{-S}. \tag{3.4}$$

Here the integral is over all trajectories $x(t)$. As it stands, eq. (3.4) is rather poorly defined. To characterize the possible trajectories we introduce a cutoff in the form of a time lattice. Putting our system into a time box of total length τ, we divide this interval into

$$N = \tau/a, \tag{3.5}$$

discrete time slices, where a is the timelike lattice spacing. Associated with

8

the i'th such slice is a coordinate x_i. This construction is sketched in figure 3.1. Replacing the time derivative of x with a nearest-neighbor difference, we reduce the action to a sum

$$S = a \sum_i \left[\tfrac{1}{2} m \left(\frac{x_{i+1} - x_i}{a} \right)^2 + V(x_i) \right]. \qquad (3.6)$$

The integral in eq. (3.4) is now defined as an integral over all the coordinates

$$Z = \int \left(\prod_i dx_i \right) e^{-S}. \qquad (3.7)$$

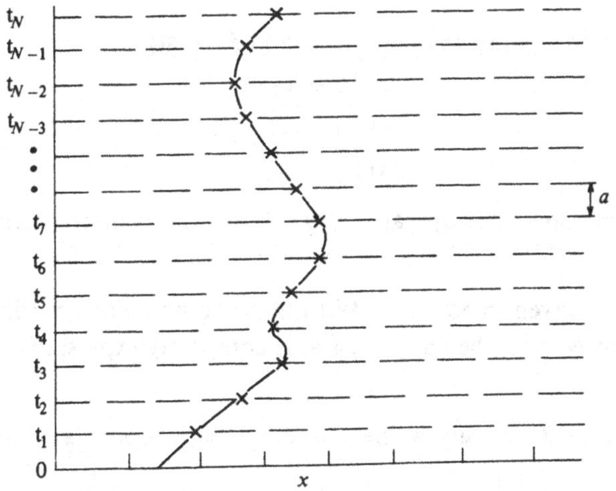

Fig. 3.1. Dividing time into a lattice. (From Creutz and Freedman, 1981.)

Eq. (3.7) is precisely in the form of a partition function for a statistical system. We have a one-dimensional chain of coordinates x_i. The action represents the inverse temperature times the Hamiltonian of the thermal analog. We will now show that evaluation of this partition function is equivalent to diagonalizing a quantum mechanical Hamiltonian obtained from this action with canonical methods. This is done via the transfer matrix.

The key to the transfer-matrix analysis is to note that the local nature of the action in eq. (3.6) permits us to write the partition function in the form of a matrix product

$$Z = \int \prod_i dx_i \, T_{x_{i+1}, x_i}, \qquad (3.8)$$

where the transfer-matrix elements are

$$T_{x', x} = \exp \left[-\frac{m}{2a} (x' - x)^2 - \frac{a}{2} (V(x') + V(x)) \right]. \qquad (3.9)$$

This operator acts in the Hilbert space of square integrable functions, where the inner product is the standard

$$\langle \psi' | \psi \rangle = \int dx \, \psi'^*(x) \, \psi(x). \qquad (3.10)$$

We introduce the non-normalizable basis states $\{|x\rangle\}$ such that

$$|\psi\rangle = \int dx \, \psi(x) |x\rangle, \qquad (3.11)$$

$$\langle x' | x \rangle = \delta(x' - x), \qquad (3.12)$$

$$1 = \int dx |x\rangle \langle x|. \qquad (3.13)$$

The canonically conjugate operators $\hat{p}$ and $\hat{x}$ satisfy

$$\hat{x} |x\rangle = x |x\rangle, \qquad (3.14)$$

$$[\hat{p}, \hat{x}] = -i, \qquad (3.15)$$

$$e^{-i\hat{p}\Delta} |x\rangle = |x + \Delta\rangle. \qquad (3.16)$$

In this Hilbert space the operator T is defined via its matrix elements

$$\langle x' | T | x \rangle = T_{x', x}, \qquad (3.17)$$

where $T_{x', x}$ is given in eq. (3.8). With periodic boundary conditions for our lattice of N sites, the path integral is compactly expressed

$$Z = \text{Tr}(T^N). \qquad (3.18)$$

The operator T is easily written in terms of the conjugate variables $\hat{p}$ and $\hat{x}$

$$T = \int d\Delta \, e^{-aV(\hat{x})/2} \, e^{-\Delta^2 m/(2a) - i\hat{p}\Delta} \, e^{-aV(\hat{x})/2}. \qquad (3.19)$$

To prove this equation, simply check that the right hand side has the matrix elements of eq. (3.9). The integral over Δ is Gaussian and gives

$$T = (2\pi a/m)^{\frac{1}{2}} e^{-\frac{1}{2}aV(\hat{x})} e^{-\frac{1}{2}a\hat{p}^2/m} e^{-\frac{1}{2}aV(\hat{x})}. \qquad (3.20)$$

Connection with the usual quantum mechanical Hamiltonian appears in the small lattice spacing limit. When a is small, the exponents in eq. (3.20) combine to give

$$T = (2\pi a/m)^{\frac{1}{2}} e^{-aH + O(a^3)}, \qquad (3.21)$$

where

$$H = \hat{p}^2/(2m) + V(\hat{x}). \qquad (3.22)$$

This is just the canonical Hamiltonian corresponding to the Lagrangian in eq. (3.1).

The procedure for going from a path-integral to a Hilbert-space formulation of quantum mechanics consists of three steps. First define the path integral with a time lattice. Then construct the transfer matrix and the Hilbert space on which it operates. Finally, take the logarithm of the transfer matrix and identify the negative of the coefficient of the linear term in the lattice spacing as the Hamiltonian. Physically, the transfer matrix propagates the system from one time to the next. Such time translations are generated by the Hamiltonian. Denoting the i'th eigenvalue of the transfer matrix by λ_i, eq. (3.18) becomes

$$Z = \sum_i \lambda_i^N. \tag{3.23}$$

As the number of time slices goes to infinity, this expression is dominated by the largest eigenvalue λ_0

$$Z = \lambda_0^N \times [1 + O(\exp[-N \ln(\lambda_0/\lambda_1)])]. \tag{3.24}$$

Thus in statistical mechanics the thermodynamic properties of a system follow from this largest eigenvalue. In ordinary quantum mechanics the corresponding eigenvector is the lowest eigenstate of the Hamiltonian; it is the ground state or, in field theory, the vacuum. Note that in this discussion the connection between imaginary and real time is trivial. Whether the generator of time translations is H or iH, we still have the same operator to diagonalize.

In statistical mechanics one is often interested in correlation functions of the statistical variables. This corresponds to a study of the Green's functions of the corresponding field theory. These are obtained upon insertion of polynomials of the fundamental variables into the path integral. We define the two-point function

$$\langle x_i x_j \rangle = (1/Z) \int \left(\prod_k dx_k \right) x_i x_j e^{-S}. \tag{3.25}$$

In terms of the transfer matrix, this reduces, for positive $i-j$, to

$$\langle x_i x_i \rangle = (1/Z) \operatorname{Tr}(T^{N-i+j} \hat{x} T^{i-j} \hat{x}). \tag{3.26}$$

Taking the length of our time box to infinity while holding the separation of i and j fixed we obtain

$$\langle x_i x_j \rangle = \langle 0 | \hat{x} (T/\lambda_0)^{i-j} \hat{x} | 0 \rangle, \tag{3.27}$$

where $|0\rangle$ is the ground state, which dominated in eq. (3.24). For a continuum limit, we hold the physical time between i and i fixed

$$t = (i-j)a, \tag{3.28}$$

and let a go to zero. We now introduce the time-dependent operator

$$\hat{x}(t) = e^{Ht} \hat{x} e^{-Ht}, \tag{3.29}$$

which corresponds to the quantum mechanical coordinate in the Heisenberg representation, but rotated to our imaginary time. Defining a time-ordering instruction to include negative time separations in the above, we identify

$$\langle x_i x_j \rangle = \langle 0|\mathcal{T}(\hat{x}(t)\,\hat{x}(0))|0\rangle$$
$$= \theta(t)\langle 0|\hat{x}(t)\,\hat{x}(0)|0\rangle + \theta(-t)\langle 0|\hat{x}(0)\,\hat{x}(t)|0\rangle, \quad (3.30)$$

where
$$\theta(t) = \begin{cases} 1, & t \geqslant 0 \\ 0, & t < 0. \end{cases} \quad (3.31)$$

This is a general result; the correlation functions of the statistical analog correspond directly with the time-ordered products of the corresponding quantum fields. It is precisely this point which allows the particle physicist to borrow technology from statistical mechanics.

In this chapter we have seen that statistical mechanics and quantum mechanics have deep mathematical connections. In general, a d-space-time dimensional quantum field theory is equivalent to a d-Euclidian dimensional classical statistical system.

Quantum statistical mechanics can also be related to quantum field theory. If we combine eqs (3.18) and (3.21) we obtain

$$Z = (2\pi a/m)^{N/2}\,\mathrm{Tr}(e^{-aNH}). \quad (3.32)$$

If we now identify
$$T = (aN)^{-1}, \quad (3.33)$$

and do not go to the large time limit, we see that a path integral in a periodic temporal box is itself a partition function at a temperature corresponding to the inverse of this periodic time. Thus the path integral formulation also enables us to study the quantum statistical mechanics of the original $(d-1)$-space dimensional theory. We will return to this point when we discuss gauge theories at finite physical temperatures and the resulting deconfining phase transitions.

Problems

1. Consider the harmonic oscillator with $V(x) = \tfrac{1}{2}kx^2$. Diagonalize the operator T of eq. (3.20). (Hint: find an operator of form $\tfrac{1}{2}p^2 + \tfrac{1}{2}\omega^2 x^2$ which commutes with T and can thus be simultaneously diagonalized.) (Creutz and Freedman, 1981.)

2. In the harmonic oscillator example, find the 'propagator' $\langle x_i x_j \rangle$.

3. Show that $a^{-2}\langle (x_{i+1}-x_i)^2 \rangle$ diverges as a goes to zero. Show that the split point product $a^{-2}\langle (x_{i+1}-x_i)(x_i-x_{i-1}) \rangle$ approaches $-\langle 0|p^2|0\rangle$ in the continuum limit. Where does the minus sign come from?

4. Calculate the fluctuations in the propagator:

$$D^2(i,j) = \langle (x_i x_j)^2 \rangle - \langle x_i x_j \rangle^2.$$

Show that the fluctuations in the split point product of problem 2 diverge as a goes to zero. Derive the virial theorem for the continuum theory:

$$\langle 0|\hat{p}^2|0\rangle = \langle 0|\hat{x}V'(\hat{x})|0\rangle.$$

This gives the average momentum squared without large fluctuations.

4
Scalar fields

The simplest quantum field theory is that of a free scalar particle. On a lattice this becomes the Gaussian model of statistical mechanics. Here we will solve this system exactly to introduce lattice field theory. As with the conventional continuum theory, Fourier transform techniques are the key to this solution. We conclude this chapter with some general remarks on interacting scalar fields.

We begin with the standard Lagrangian density for a self-conjugate free field

$$\mathcal{L} = \tfrac{1}{2}(\partial_\mu \phi)^2 + \tfrac{1}{2}m^2\phi^2. \tag{4.1}$$

Here $\phi(x)$ is a real function of the four space-time coordinates x_μ. The discussion here is easily generalized to an arbitrary number of dimensions and complex fields. The Greek indices denoting vector quantities run from one to four. A repeated index, as implied in eq. (4.1), is understood to be summed; however, as we work in Euclidian space, no metric tensor is implied. To every field configuration corresponds an action

$$S = \int \mathrm{d}^4x\, \mathcal{L}. \tag{4.2}$$

The Feynman path integral is a sum over all configurations

$$Z = \int [\mathrm{d}\phi]\, e^{-S}, \tag{4.3}$$

where, as in the previous chapter, the integration measure needs definition.

We proceed directly to a four-dimensional hypercubic lattice. Thus we restrict our coordinates to the form

$$x_\mu = a n_\mu, \tag{4.4}$$

where a is the lattice spacing and n_μ has four integer components. As an infrared cutoff, we allow the individual components of n to assume only a finite number N of independent values

$$-N/2 < n_\mu \leqslant N/2. \tag{4.5}$$

Outside this range we assume the lattice is periodic; we identify n with $n+N$. Thus our lattice has N^4 sites. We now replace the derivatives of ϕ with nearest neighbor differences

$$\partial_\mu \phi(x_\nu) \to (\phi_{n_\nu + \delta_{\nu\mu}} - \phi_{n_\nu})/a, \tag{4.6}$$

where the Kronecker function is defined

$$\delta_{\mu\nu} = \begin{cases} 1, & \mu = \nu \\ 0, & \mu \neq \nu. \end{cases} \tag{4.7}$$

The action is a sum

$$S = a^4 [\sum_{\{m,n\}} (\phi_m - \phi_n)^2/(2a^2) + \sum_n m^2\phi_n^2/2], \tag{4.8}$$

where $\{m, n\}$ represents the set of all nearest-neighbor pairs of lattice sites. The path integration measure is now simply defined as an ordinary integral over each of the lattice fields

$$Z = \int (\prod_n d\phi_n) e^{-S}. \tag{4.9}$$

At this point we observe that the action is a quadratic form in the field variables

$$S = \tfrac{1}{2}\phi_m M_{mn} \phi_n, \tag{4.10}$$

where M is an N^4-dimensional square matrix and we adopt the usual summation convention on repeated indices. The integral in eq. (4.9) is of the standard Gaussian form and has the value

$$Z = |M/2\pi|^{-\frac{1}{2}}, \tag{4.11}$$

where the vertical bars denote the determinant of the enclosed matrix. We will now introduce a Fourier transform on the lattice. This will diagonalize M and make the determinant trivial.

Let f_n be an arbitrary complex function on the lattice sites. Its Fourier transform is defined

$$\tilde{f}_k = F_{kn}f_n = \sum_n f_n e^{2\pi i k_\mu n_\mu/N}. \tag{4.12}$$

The index k also carries four integer valued components, each in the range of eq. (4.5). This linear transform is easily inverted with the identity

$$\sum_k e^{-2\pi i k \cdot n/N} = N^4 \prod_\mu \delta_{n_\mu,0} \equiv N^4 \delta_{n,0}^4. \tag{4.13}$$

Thus we have $\quad (F^{-1})_{nk} = N^{-4} e^{-2\pi i k \cdot n/N} = N^{-4} F_{kn}^* \tag{4.14}$

or $\quad f_n = N^{-4} \sum_k \tilde{f}_k e^{-2\pi i k \cdot n/N}. \tag{4.15}$

The utility of the Fourier series appears when we consider sums of local quadratic forms, such as appear in our lattice action. In particular, the useful identities

$$\sum_n f_n^* g_n = N^{-4} \sum_k \tilde{f}_k^* \tilde{g}_k \tag{4.16}$$

and $\quad \sum_n f_{n_\mu+\delta_{\mu\nu}}^* g_n = N^{-4} \sum_k \tilde{f}_k^* \tilde{g}_k e^{2\pi i k_\nu/N} \tag{4.17}$

reduce the action to $\quad S = a^4 N^{-4} \sum_k \tfrac{1}{2}\tilde{M}_k |\tilde{\phi}_k|^2, \tag{4.18}$

where
$$\tilde{M}_k = m^2 + 2a^{-2}\sum_{\mu}(1 - \cos(2\pi k_{\mu}/N)). \tag{4.19}$$

The Fourier transform has diagonalized M

$$M_{mn} = a^4 N^{-4}\sum_{k}F^*_{mk}F_{nk}\tilde{M}_k. \tag{4.20}$$

To evaluate the determinant of this matrix, first note that eq. (4.14) implies

$$|N^{-4}F^*| = |F|^{-1}. \tag{4.21}$$

Thus we have the exact expression for our path integral

$$Z = |M/2\pi|^{-\frac{1}{2}} = \prod_{k}(a^4\tilde{M}_k/2\pi)^{-\frac{1}{2}}. \tag{4.22}$$

This equation is not very useful as it stands. To obtain Green's functions, we consider external sources J_n on the lattice sites and coupled to the field ϕ. Consequently we generalize our action to

$$S(J) = \tfrac{1}{2}\phi_m M_{mn}\phi_n - J_n\phi_n. \tag{4.23}$$

The partition function now depends on the sources

$$Z(J) = \int[\mathrm{d}\phi]\,e^{-S(J)}. \tag{4.24}$$

This quantity is a generating function for the Green's functions, which follow from differentiation with respect to the sources

$$\langle\phi_{n_1}\cdots\phi_{n_j}\rangle = Z^{-1}\int[\mathrm{d}\phi]\,e^{-S}\phi_{n_1}\cdots\phi_{n_j}\bigg|_{J=0}$$
$$= Z^{-1}\left\{\frac{\mathrm{d}}{\mathrm{d}J_{n_1}}\cdots\frac{\mathrm{d}}{\mathrm{d}J_{n_j}}Z(J)\right\}\bigg|_{J=0} \tag{4.25}$$

Completing the square in eq. (4.23) and shifting the integration in eq. (4.24) gives the exact expression for this free-field generating function

$$Z(J) = Z(0)\exp\left(\tfrac{1}{2}J_m(M^{-1})_{mn}J_n\right), \tag{4.26}$$

where $Z(0)$ is given in eq. (4.22). From this we see that the propagator or two-point function is simply the inverse of the matrix M

$$\langle\phi_m\phi_n\rangle = (M^{-1})_{mn}. \tag{4.27}$$

Momentum space makes this inversion trivial

$$\langle\phi_m\phi_n\rangle = a^{-4}N^{-4}\sum_{k}\tilde{M}_k^{-1}e^{2\pi i k\cdot(m-n)/N}. \tag{4.28}$$

To put this expression into a more familiar form, we first take N to infinity and change the momentum sum into an integral with the replacements

$$q_{\mu} = 2\pi k_{\mu}/(Na), \tag{4.29}$$

$$a^{-4}N^{-4}\sum_{k} \to \int\frac{\mathrm{d}^4q}{(2\pi)^4}. \tag{4.30}$$

Here each component of q runs over the finite range

$$-\pi/a < q_\mu \leqslant \pi/a. \tag{4.31}$$

This explicitly shows the momentum space effect of the lattice cutoff. The propagator now assumes the form

$$\langle \phi_m \phi_n \rangle = \int_{-\pi/a}^{\pi/a} \frac{\mathrm{d}^4 q}{(2\pi)^4} \frac{e^{-iq \cdot x}}{m^2 + 2a^{-2} \sum_\mu [1 - \cos(aq_\mu)]}, \tag{4.32}$$

where

$$x_\mu = -a(m_\mu - n_\mu). \tag{4.33}$$

For the continuum limit $a \to 0$ we expand the cosine

$$2a^{-2} \sum_\mu (1 - \cos(aq_\mu)) = q^2 + O(a^2) \tag{4.34}$$

and obtain

$$\langle \phi_m \phi_n \rangle = \int \frac{\mathrm{d}^4 q}{(2\pi)^4} \frac{e^{-iq \cdot x}}{m^2 + q^2} + O(a^2). \tag{4.35}$$

This is the familiar Feynman propagator function in Euclidian space.

Up to this point we have been considering a free field. Now we add an interaction term to our action

$$S = \tfrac{1}{2}\phi_m M_{mn} \phi_n - J_n \phi_n + \sum_n V_I(\phi_n). \tag{4.36}$$

The full potential felt by the field ϕ includes the mass term from eq. (4.8)

$$V(\phi) = \tfrac{1}{2}m^2 \phi^2 + V_I(\phi). \tag{4.37}$$

The minima of this function form the basis for semiclassical treatments, with which we will not concern ourselves here. As a concrete example, the usual ϕ^4 theory takes

$$V_I(\phi) = g_0 \phi^4. \tag{4.38}$$

Here g_0 is the bare coupling with the lattice cutoff in place. The full generating function of the interacting theory is still

$$Z(J) = \int [\mathrm{d}\phi] e^{-S(J)}. \tag{4.39}$$

Note that the potential $V(\phi)$ must be bounded below if this integral is to make any sense. In particular, the ϕ^4 theory with negative coupling is sick, and therefore we do not expect analyticity at vanishing g_0. Perturbation in g_0 yields at best an asymptotic series (Dyson, 1952).

The usual perturbation expansion follows from a formal exploitation of eq. (4.25) to give

$$Z(J) = \exp\left(\sum_n V_I(\mathrm{d}/\mathrm{d}J_n)\right) Z_0(J), \tag{4.40}$$

where $Z_0(J)$ is the free-field generating function from eq. (4.26). An expansion of the exponent in this equation gives the Feynman series in terms of vertices from the interaction term and propagators from $Z_0(J)$.

The Green's functions, which follow by differentiating Z with respect to the sources, are the full n-point functions and include, in general, disconnected pieces. In particular, if ϕ has a vacuum expectation value, one might prefer to subtract this and study the connected propagator

$$\langle \phi_m \phi_n \rangle_c = \langle \phi_m \phi_n \rangle - \langle \phi_m \rangle \langle \phi_n \rangle. \tag{4.41}$$

A general connected Green's function is defined through the corresponding generating function, which is simply the logarithm of Z

$$F(J) = \ln(Z(J)), \tag{4.42}$$

$$\langle \phi_{n_1} \cdots \phi_{n_j} \rangle_c = \left(\frac{\mathrm{d}}{\mathrm{d}J_{n_1}} \cdots \frac{\mathrm{d}}{\mathrm{d}J_{n_j}} F(J) \right) \Bigg|_{J=0}. \tag{4.43}$$

Note that in the statistical mechanical analog $F(J)$ is proportional to the free energy.

We conclude this chapter with some brief remarks on the strong coupling expansion for this scalar theory. Considering the ϕ^4 theory of eq. (4.38), we change integration variables in the path integral from ϕ to $g_0^{\frac{1}{4}}\phi$, and we perform a similar formal manipulation to that giving eq. (4.40). Thus we find

$$Z(g_0^{\frac{1}{4}}J) = g_0^{-N^4/4} \int [\mathrm{d}\phi] \, \mathrm{e}^{-\frac{1}{2}g_0^{-\frac{1}{2}}\phi M \phi} \mathrm{e}^{-\Sigma_n(\phi_n^4 - J_n\phi_n)}$$

$$= g_0^{-N^4/4} \exp\left(-\frac{1}{2}g_0^{-\frac{1}{2}} \frac{\mathrm{d}}{\mathrm{d}J} M \frac{\mathrm{d}}{\mathrm{d}J} \right) \prod_n f(J_n), \tag{4.44}$$

where $f(J)$ is an ordinary one-dimensional integral

$$f(J) = \int_{-\infty}^{\infty} \mathrm{d}\phi \, \mathrm{e}^{-(\phi^4 - J\phi)}. \tag{4.45}$$

An expansion of the exponential on the right hand side of eq. (4.44) forms the basis for a strong coupling expansion in powers of $g_0^{-\frac{1}{2}}$. Unfortunately, in the continuum limit the matrix M grows, and therefore for fixed coupling we are no longer expanding in a small quantity. As we are more interested in gauge theories, we will not discuss here the techniques invented in attempts to overcome this problem. We only wish to emphasize that a strong coupling series is quite natural when the lattice is in place (Baker and Kincaid, 1979; Bender *et al.*, 1981).

Problems

1. Verify equation (4.19).

2. Show that a rescaling of the field normalization puts the action in the form

$$S = \sum_m \phi_m^2 + K \sum_{\{mn\}} \phi_m \phi_n.$$

Show that in the continuum limit the 'hopping constant' K goes to unity at a rate dependent on the mass.

3. One might consider as a non-perturbative cutoff disregarding a field's Fourier components which carry momentum larger than some cutoff parameter. How does this compare to the lattice cutoff in real space?

5

Fermions

In this chapter we turn to a subject that is still not completely understood, the lattice formulation of fermionic fields. These complications with spinor particles are already present at the free-field level; a straightforward generalization of the ideas in the last chapter does not give a simple particle spectrum. The action needs additional terms which vanish in the naive continuum limit. These terms, needed to eliminate certain lattice artifacts, tend to mutilate the classical symmetries of the theory. The extent to which this is necessary is still an open question.

Before proceeding to these topics, we must introduce the concepts of anticommuting numbers and integration over these variables. The path integral is no longer a sum, but a particular linear operation from functions of anticommuting variables into the complex numbers. We will also introduce anticommuting sources for the dynamical fields. A differentiation with respect to these sources gives the Green's functions, as in the last chapter. Both integration and differentiation with anticommuting variables have useful analogous properties to ordinary integrals and derivatives; however, there are some amusing distinctions. In particular, fermionic integrals and derivatives involve essentially the same operation.

As in the previous chapter, we begin with the continuum Lagrangian density for a free field, in this case a four-component Dirac spinor

$$\mathcal{L} = \bar{\psi}(\slashed{\partial} + m)\,\psi. \tag{5.1}$$

A slash through a four-vector represents the usual sum

$$\slashed{p} = p_\mu \gamma_\mu, \tag{5.2}$$

where the γ_μ are a set of four-by-four Euclidian Dirac matrices satisfying the algebra

$$[\gamma_\mu, \gamma_\nu]_+ = \gamma_\mu \gamma_\nu + \gamma_\nu \gamma_\mu = 2\delta_{\mu\nu} \tag{5.3}$$

$$\gamma_\mu^\dagger = \gamma_\mu. \tag{5.4}$$

As usual, we define
$$\bar{\psi} = \psi^\dagger \gamma_4. \tag{5.5}$$

For future use we also introduce

$$\gamma_5 = \gamma_1 \gamma_2 \gamma_3 \gamma_4 = \gamma_5^\dagger. \tag{5.6}$$

A convenient representation for these matrices is

$$\gamma_i = \begin{pmatrix} 0 & \sigma_i \\ \sigma_i & 0 \end{pmatrix} \quad i = 1, 2, 3, \tag{5.7}$$

$$\gamma_4 = \begin{pmatrix} 1 & 0 \\ 0 & -1 \end{pmatrix}, \tag{5.8}$$

$$\gamma_5 = \begin{pmatrix} 0 & -i \\ i & 0 \end{pmatrix}. \tag{5.9}$$

The matrix elements here are themselves two-by-two matrices and the σ_i are the usual Pauli matrices

$$\sigma_1 = \begin{pmatrix} 0 & 1 \\ 1 & 0 \end{pmatrix}, \tag{5.10}$$

$$\sigma_2 = \begin{pmatrix} 0 & -i \\ i & 0 \end{pmatrix}, \tag{5.11}$$

$$\sigma_3 = \begin{pmatrix} 1 & 0 \\ 0 & -1 \end{pmatrix}. \tag{5.12}$$

Note that this Lagrangian is invariant under the substitution

$$\psi \to e^{i\theta}\psi. \tag{5.13}$$

This symmetry is directly related to the conservation of fermion number. When the mass m vanishes, the theory also has a 'chiral' or 'γ_5' symmetry under

$$\psi \to e^{i\theta\gamma_5}\psi. \tag{5.14}$$

In a naive canonical treatment, these symmetries are generated by the currents

$$j_\mu = \bar{\psi}\gamma_\mu\psi \tag{5.15}$$

and

$$j_\mu^5 = \bar{\psi}\gamma_\mu\gamma_5\psi. \tag{5.16}$$

Careful perturbative analysis (Adler, 1969; Bell and Jackiw, 1969) indicates the impossibility of maintaining conservation of both these currents in the four-dimensional quantum theory. This 'anomaly' will not be derived here; we only note that it is deeply related to the difficulties encountered in the lattice formulation, which naively preserves these symmetries (Chodos and Healy, 1977; Nielsen and Ninomiya, 1981a, b; Kerler, 1981a; Becher and Joos, 1982; Rabin, 1982).

As in the previous chapter, we introduce a four-dimensional hypercubic lattice of N^4 sites. With each site m we associate an independent four-component spinor variable ψ_m. To keep the lattice action simple we define the derivative symmetrically

$$\partial_\mu\psi \to \frac{1}{2a}(\psi_{m_\nu + \delta_{\mu\nu}} - \psi_{m_\nu - \delta_{\mu\nu}}). \tag{5.17}$$

Summing the Lagrangian over all sites gives the lattice action

$$S = \sum_{m,n} \bar{\psi}_m M_{mn} \psi_n, \tag{5.18}$$

where $\quad M_{mn} = \tfrac{1}{2} a^3 \sum_{\mu} \gamma_\mu (\delta^4_{m_\nu + \delta_{\mu\nu},\, n_\nu} - \delta^4_{m_\nu - \delta_{\mu\nu},\, n_\nu}) + a^4 m \delta^4_{mn}. \tag{5.19}$

Note that the symmetries of eq. (5.13) and, when $m = 0$, eq. (5.14) are still manifest. We now put this action into a path integral

$$Z_0 = \int [d\psi \, d\bar{\psi}] \, e^{-S}. \tag{5.20}$$

Unlike in the scalar case, this is not an ordinary integral, and needs further definition. We will first discuss such integrals for quadratic actions of the form of eq. (5.18) with an arbitrary matrix M. Later we will return to the specific theory in eq. (5.19).

We begin by requiring the integration variables to anticommute

$$[\psi^\alpha_m, \psi^\beta_n]_+ = [\psi^{\alpha\dagger}_m, \psi^\beta_n]_+ = [\psi^{\alpha\dagger}_m, \psi^{\beta\dagger}_n]_+ = 0, \tag{5.21}$$

where α and β are the usually suppressed spinor indices. This equation contrasts sharply with the canonical relations for Dirac operators in Hilbert space. In the path integral ψ and $\psi^\dagger$ are independent fermionic objects. As in the previous chapter, our integral is of an exponentiated quadratic form. We will see that its evaluation again reduces to knowing the determinant of M. Before proceeding, however, we find it advantageous at this point to introduce the concept of sources for the fermionic fields.

As our fields anticommute, any sources coupled to them should behave similarly. We consider separate sources b^α_m and c^α_m for ψ^α_m and $\bar{\psi}^\alpha_m$, respectively. Suppressing repeated site and spinor indices, we generalize the action to

$$S = \bar{\psi} M \psi + b\psi - \bar{\psi} c. \tag{5.22}$$

We adopt the convention that all the spinor quantities ψ, $\bar{\psi}$, b and c anticommute with themselves and each other. We wish to define the fermionic path integral such that the linear source terms can be eliminated by a simple completion of the square and a shift of the integration variables, in analogy to an ordinary integral. Thus we demand

$$Z = Z_0 \exp(-bM^{-1}c), \tag{5.23}$$

where Z_0 is the sourceless integral from eq. (5.20). For the free field considered here, the overall factor of Z_0 is irrelevant to the evaluation of Green's functions. In particular, the fermion propagator is given, as for scalar fields, by the inverse of the kinetic matrix M. However, in more general applications, i.e. with gauge fields, one may wish to have M to depend on other interacting fields. In this case we need the explicit functional dependence of Z_0 on this matrix. We will now demonstrate that Z_0 is simply the determinant of M (Matthews and Salam, 1954).

To proceed, we need the concept of derivatives with respect to our fermionic sources. Such derivatives should satisfy

$$\left[\frac{d}{db_m^\alpha}, b_n^\beta\right]_+ = \delta_{mn}\delta^{\alpha\beta} \tag{5.24}$$

and a corresponding equation for the c's. This generalizes ordinary differentiation, where one would have a commutator. Note that these anticommutation relations are precisely those of the creation and annihilation operators for fermions on the sites of our lattice

$$[b_m^{\alpha\dagger}, b_n^\beta]_+ = \delta_{mn}\delta^{\alpha\beta}. \tag{5.25}$$

We can realize these relations on a Fock space of states created by

$$\frac{d}{db_m^\alpha} \leftrightarrow b_m^{\alpha\dagger} \tag{5.26}$$

on a 'vacuum' satisfying

$$b_m^\alpha|0\rangle = c_m^\alpha|0\rangle = 0. \tag{5.27}$$

Operating on this vacuum is equivalent to turning off the sources. This 'Euclidian vacuum' should not be confused with the conventional Hilbert space state in Minkowski space, as found in the transfer matrix formalism discussed in the first chapter. We use a creation operator notation here for compactness and to avoid confusion from the fact that d/db is not really a usual derivative.

Our path integral with sources is an operator in this space. Operating on the vacuum from the right, we define the useful state

$$\langle Z| = \langle 0|Z(b, c). \tag{5.28}$$

Fermion Green's functions are matrix elements between this state and the vacuum

$$\int [d\psi\, d\bar{\psi}]\, e^{-S} \bar{\psi}_{i_1} \dots \bar{\psi}_{i_n} \psi_{j_1} \dots \psi_{j_n} = \langle Z|c_{i_1}^\dagger \dots c_{i_n}^\dagger b_{j_1}^\dagger \dots b_{j_n}^\dagger|0\rangle. \tag{5.29}$$

Our creation operators produce the ends of the external lines in a general correlation function.

Continuing toward our goal of evaluating Z_0, we now present a useful identity on exponentials of quadratic forms in creation and annihilation operators. Let F and G be N^4-by-N^4 symmetric matrices. We would like to take the expression

$$\langle\psi(\lambda)| = \langle 0|e^{bFc}e^{\lambda b^\dagger Gc^\dagger} \tag{5.30}$$

and manipulate the creation operators to the left to obtain a single exponential of a quadratic form in the annihilation operators alone.

Straightforward manipulations, which we invite the reader to perform, yield the identities

$$(\psi(\lambda)|b^\dagger = -(\psi(\lambda)|(F^{-1}-\lambda G)^{-1}c, \qquad (5.31)$$

$$(\psi(\lambda)|c^\dagger = +(\psi(\lambda)|(F^{-1}-\lambda G)^{-1}b. \qquad (5.32)$$

Using these in the derivative of expression (5.30) with respect to the parameter λ gives a differential equation

$$\frac{d}{d\lambda}(\psi| = (\psi|[-\text{Tr}(G(F^{-1}-\lambda G)^{-1})+b(F^{-1}-\lambda G)^{-1}G(F^{-1}-\lambda G)^{-1}c].$$

$$(5.33)$$

With the initial condition

$$(\psi(\lambda = 0)| = (0|e^{bFc}, \qquad (5.34)$$

we can integrate to obtain

$$(\psi| = |I-\lambda FG|(0|\exp{(b(F^{-1}-\lambda G)^{-1}c)}. \qquad (5.35)$$

To verify that this is indeed a solution of eq. (5.33), make use of the well-known identity

$$|F| = \exp{[\text{Tr}(\ln F)]}. \qquad (5.36)$$

With eq. (5.33) in hand, we return to our path integral and write

$$M = I+(M-I), \qquad (5.37)$$

where I is the identity matrix. Treating $M-I$ as a perturbation, we have

$$(Z| = (0|\int[d\psi\,d\bar{\psi}]e^{-\bar{\psi}\psi-b\psi+\bar{\psi}c}\exp{(c^\dagger(I-M)b^\dagger)}. \qquad (5.38)$$

As before, the integral can be done by completing the square; however, now the normalization is truly arbitrary. We define

$$\int[d\psi\,d\bar{\psi}]e^{-\bar{\psi}\psi} = 1. \qquad (5.39)$$

Thus we have

$$(Z| = (0|e^{-bc}e^{-b^\dagger(I-M)c^\dagger}. \qquad (5.40)$$

This is exactly in the form needed for the identity in eq. (5.33), which gives the final result

$$(Z| = |M|(0|e^{-bM^{-1}c}. \qquad (5.41)$$

Turning off the sources, we see that Z_0 is precisely the determinant of M

$$|M| = \int[d\psi\,d\bar{\psi}]e^{-\bar{\psi}M\psi}. \qquad (5.42)$$

Note the similarity of this with the boson result in eq. (4.22). The anticommuting fermion fields have moved the determinant from the denominator to the numerator. For scalar fields there is also an operator formalism parallel to that presented here. We invite the reader to work out the Bose analog of eq. (5.35).

This discussion paid no attention to the precise form of the kinetic matrix

M; we only used the quadratic nature of the fermion action. We now return to the specific case in eq. (5.19) and study the propagator

$$S_{mn} = Z_0^{-1} \int [\mathrm{d}\psi \, \mathrm{d}\bar\psi] \, e^{-\bar\psi M \psi} \, \psi_m \bar\psi_n. \qquad (5.43)$$

In our operator formalism we have

$$S_{mn} = Z_0^{-1}(Z|b_m^\dagger c_n^\dagger|0) = (M^{-1})_{mn}. \qquad (5.44)$$

As in the last chapter, M is diagonalized and inverted with a Fourier transform. This gives

$$(M^{-1})_{mn} = a^{-4}N^{-4} \sum_k \tilde{M}_k^{-1} e^{2\pi i k \cdot (m-n)/N}, \qquad (5.45)$$

where
$$\tilde{M}_k = m + i a^{-1} \sum_\mu \gamma_\mu \sin(2\pi k/N). \qquad (5.46)$$

Still following the last chapter, we let our lattice size become large and replace sums over k with integrals

$$q_\mu = 2\pi k_\mu/(Na), \qquad (5.47)$$

$$a^{-4}N^{-4} \sum_k \rightarrow \int \mathrm{d}^4 q/(2\pi)^4, \qquad (5.48)$$

$$\tilde{M}_k = m + i a^{-1} \sum_\mu \gamma_\mu \sin(aq_\mu). \qquad (5.49)$$

If we now consider small lattice spacing and expand in powers of a, we find
$$\tilde{M}_k = m + i\slashed{q} + O(a^2). \qquad (5.50)$$

It thus appears that we have recovered the usual continuum fermion propagator. Unfortunately, more care is needed at the upper limits of the momentum integrals. When q_μ is π/a, the periodic sine function in eq. (5.49) vanishes. Here the $O(a^2)$ terms cannot be neglected. Indeed, the propagator has no supression of momentum values near π/a; therefore we must expect rapid variations in the fields from site to neighboring site. This precludes the above simple continuum limit and will also destroy any attempt to formulate a transfer matrix along the lines of chapter 3.

To isolate the large momentum region, consider one component of q_μ and replace it with
$$\tilde{q}_\mu = q_\mu - \pi/a \qquad (5.51)$$

over half the integration region

$$\int_{-\pi/a}^{\pi/a} \mathrm{d}q_\mu \, \tilde{M}_k^{-1} = \int_{-\pi/2a}^{\pi/2a} (\mathrm{d}q_\mu + \mathrm{d}\tilde{q}_\mu) \, \tilde{M}_k^{-1}. \qquad (5.52)$$

For small lattice spacing, a finite range of the integration variables dominates each of these terms. Now an approximation along the lines of eq. (5.50) is valid. For each space-time dimension we have two independent regions where the theory gives a free fermion propagator in the continuum limit. We actually have $2^4 = 16$ independent fermion species, even though we initially seemed to have but one.

This multiplicity in the spectrum arose because we have implemented a regularization scheme that, when $m = 0$, keeps an exact γ_5 symmetry at all stages. It therefore cannot possess the known chiral anomaly. The theory has created extra species which cancel this phenomenon. Note that the new fermions use different γ matrices; i.e. when we shift as in eq. (5.51), the sine function gives an extra negative sign

$$\gamma_\mu \sin (q_\mu a) = -\gamma_\mu \sin (\tilde{q}_\mu a). \tag{5.53}$$

This sign is absorbed by redefining γ_μ and therefore γ_5 as well. Those fermions associated with an odd number of components of q being shifted by π/a will transform under the conjugate of the rotation in eq. (5.14). We have equal numbers of states with each chirality.

Several solutions exist for this 'doubling' problem. Perhaps the simplest is to ignore it and say that the theory is automatically generating a large number of fermion 'flavors.' Indeed, real quarks do appear to come in several species. Nonetheless, it seems a bit far fetched to use an artifact of the lattice formulation to explain this degeneracy.

Observing that the problem only occurs when the magnitude of q is large, one might try artificially to exclude large components. In general this is dangerous because of completeness in the Fourier transform. Here, however, we can use the spinor index to partially do precisely this. By associating only a single spinor component with each site and putting different components on separate classes of sites, one effectively puts the components on smaller sublattices. This reduces the effective upper limit of the momentum integrals and thereby reduces some of the unwanted degeneracy. Such techniques have had considerable success in a Hamiltonian formulation of the lattice theory, where continuous time removes half of the unwanted states (Kogut and Susskind, 1975; Banks *et al.*, 1977).

The multiplicity problem arises from the periodic nature of the sine function appearing in the Fourier transform of the nearest-neighbor form for the lattice derivative. In a continuum theory, a derivative is simply a factor of the momentum in Fourier space. Thus another solution to the lattice degeneracy is to replace the sin of the momentum with the momentum itself. This defines a new lattice derivative which immediately kills the extra states. On returning to position space, this derivative no longer involves just nearby sites, but includes products of site variables with arbitrary separations. This keeps an apparent chiral symmetry; to see the anomaly requires a careful and somewhat controversial treatment of limits (Drell, Weinstein and Yankielowicz, 1976; Karsten and Smit, 1981).

A utilitarian approach to the doubling problem is to add to the naive action new terms which suppress the extra states while vanishing in a continuum limit with the desired fermion species. To keep the action as local as possible, we require the new terms to involve nearest-neighbor pairs of lattice sites. This means that in momentum space these terms will involve only simple trigonometric functions of the momentum. An addition which accomplishes our needs replaces $\tilde{M}_k$ with

$$\tilde{M}_k = m + ia^{-1}\sum_\mu \gamma_\mu \sin(aq_\mu) + ra^{-1}\sum_\mu (1-\cos(aq_\mu)). \qquad (5.54)$$

Here r is an arbitrary parameter. Note that for small momentum the new term is of order the cutoff and thus drops out. However, when a component of q is near π/a, the addition increases the mass of the unwanted state by $2r/a$. In the continuum limit all the extra states go to infinite mass and only one species of mass m survives. Setting r to unity (Wilson, 1977), we obtain the position space form

$$M_{mn} = (a^4 m + 4a^3)\delta^4_{mn}$$
$$+ \tfrac{1}{2}a^3\sum_\mu [(1+\gamma_\mu)\delta^4_{m_\nu + \delta_{\mu\nu}, n_\nu} + (1-\gamma_\mu)\delta^4_{m_\nu - \delta_{\mu\nu}, n_\nu}]. \qquad (5.55)$$

Whenever a quark moves from one site to the next, its wave function picks up a factor of $1 \pm \gamma_\mu$ rather than the γ_μ from eq. (5.19). Note that $(1 \pm \gamma_\mu)/2$ is a rank two projection

$$(\tfrac{1}{2}(1 \pm \gamma_\mu))^2 = \tfrac{1}{2}(1 \pm \gamma_\mu), \qquad (5.56)$$

$$\mathrm{Tr}\,(\tfrac{1}{2}(1 \pm \gamma_\mu)) = 2. \qquad (5.57)$$

Thus part of the spinor field no longer propagates. This reduces the degeneracy by a factor of two for each dimension, exactly as needed to remove the extra states. This method is referred to as the projection operator technique of Wilson.

The simplicity of this method is convenient for calculation. However, it totally mutilates the chiral symmetry of the theory because the added piece is like a mass term for the unwanted fermions. This is probably more of a mutilation than necessary; with several flavors not all currents need to have an anomaly. Consequences of the related symmetries, such as Goldstone bosons, are masked in the projection operator formalism until one reaches the continuum. The extent to which these latent symmetries can survive in a lattice theory is still unclear.

Problems

1. Derive the analog of eq. (5.35) for bosonic operators.
2. For a single pair of fermionic variables ψ and $\bar{\psi}$, derive the formulae

$$\int d\psi \, d\bar{\psi} \, 1 = \int d\psi \, d\bar{\psi} \, \psi =$$

$$\int d\psi \, d\bar{\psi} \, \bar{\psi} = 0$$

$$\int d\psi \, d\bar{\psi} \, \psi\bar{\psi} = 1.$$

3. Rescale the fields to put eq. (5.55) in the form

$$M_{mn} = \delta^4_{mn} + K\sum_{\mu} ((1+\gamma_\mu) \, \delta^4_{m_\nu + \delta_{\mu\nu}, \, n_\nu} + (1-\gamma_\mu) \, \delta^4_{m_\nu - \delta_{\mu\nu}, \, n_\nu}),$$

where the 'hopping constant' K approaches $1/8$ for a continuum limit. This represents a critical point where the correlation length diverges when expressed in units of the lattice spacing.

4. We have discussed periodic boundary conditions. Is there any motivation for antiperiodic boundary conditions for fermionic fields? (Hint: what sign does a fermion loop wrapping around the lattice give to eq. (3.32)?)

6

Gauge fields

What is a gauge theory? This question may have more answers than there are physicists. In this discursive chapter we digress into discussing a few general definitions of a gauge theory and, in particular, a non-Abelian theory.

At the simplest level, a non-Abelian gauge theory is merely an embellishment of electromagnetism with an internal symmetry. Electromagnetic fields form the components of an antisymmetric tensor which is a four-dimensional curl of a vector potential

$$F_{\mu\nu} = \partial_\mu A_\nu - \partial_\nu A_\mu. \tag{6.1}$$

Yang and Mills (1954) proposed adding an isospin index to A_μ and $F_{\mu\nu}$

$$A_\mu \rightarrow A_\mu^\alpha, \tag{6.2}$$

$$F_{\mu\nu} \rightarrow F_{\mu\nu}^\alpha. \tag{6.3}$$

This trivial modification becomes not so trivial with the addition of a further antisymmetric piece to $F_{\mu\nu}$

$$F_{\mu\nu}^\alpha = \partial_\mu A_\nu^\alpha - \partial_\nu A_\mu^\alpha + g_0 f^{\alpha\beta\gamma} A_\mu^\beta A_\nu^\gamma. \tag{6.4}$$

Here g_0 is the bare gauge coupling constant, and $f^{\alpha\beta\gamma}$ are the structure constants for some continuous group G.

We consider here only unitary groups. An element g of G is a matrix in the fundamental or defining representation. We parametrize the elements of G using a set of generators

$$g = e^{i\omega^\alpha \lambda^\alpha}. \tag{6.5}$$

Here the ω^α are parameters and the λ^α are a set of Hermitian matrices which generate the group. The structure constants are defined from the commutation relations

$$[\lambda^\alpha, \lambda^\beta] = i f^{\alpha\beta\gamma} \lambda^\gamma. \tag{6.6}$$

The generators are conventionally orthonormalized such that

$$\mathrm{Tr}(\lambda^\alpha \lambda^\beta) = \tfrac{1}{2}\delta^{\alpha\beta}. \tag{6.7}$$

The simplest non-Abelian theory uses the group $SU(2)$ which is generated by the Pauli matrices (eqs 5.10–12)

$$\lambda^\alpha = \tfrac{1}{2}\sigma^\alpha, \tag{6.8}$$

$$f^{\alpha\beta\gamma} = \epsilon^{\alpha\beta\gamma}. \tag{6.9}$$

Maxwell's equations for electrodynamics follow from the Lagrangian density

$$\mathcal{L} = \tfrac{1}{4}F_{\mu\nu}\,F_{\mu\nu} + j_\mu\,A_\mu. \qquad (6.10)$$

Here j_μ represents an external source for the electromagnetic field. The non-Abelian theory begins with the same Lagrangian, except for an assumed sum over a suppressed isospin index and $F_{\mu\nu}$ includes the extra term in eq. (6.4). The classical equation of motion for electromagnetism

$$\partial_\mu F_{\mu\nu} = j_\nu, \qquad (6.11)$$

picks up an extra piece in the non-Abelian theory and becomes

$$(D_\mu F_{\mu\nu})^\alpha = j_\nu^\alpha. \qquad (6.12)$$

Here the 'covariant derivative' is defined

$$(D_\mu F_{\mu\nu})^\alpha = \partial_\mu F_{\mu\nu}^\alpha + g_0 f^{\alpha\beta\gamma} A_\mu^\beta F_{\mu\nu}^\gamma. \qquad (6.13)$$

The motivation for this definition will become clear when we discuss gauge transformations. The antisymmetry of $F_{\mu\nu}$ requires that the source satisfy

$$(D_\mu j_\mu)^\alpha = 0. \qquad (6.14)$$

This is the non-Abelian analog of current conservation.

A convenient notation follows from using the group generators to define a matrix potential

$$A_\mu = A_\mu^\alpha \lambda^\alpha. \qquad (6.15)$$

Using eq. (6.7) we can invert this relation

$$A_\mu^\alpha = 2\,\mathrm{Tr}\,(\lambda^\alpha A_\mu). \qquad (6.16)$$

Similarly we define matrices for $F_{\mu\nu}$ and j_μ. The expression for $F_{\mu\nu}$ in terms of A_μ takes the simple form

$$F_{\mu\nu} = \partial_\mu A_\nu - \partial_\nu A_\mu - \mathrm{i}g_0\,[A_\mu, A_\nu]. \qquad (6.17)$$

In this notation the Lagrangian density becomes

$$\mathcal{L} = \tfrac{1}{2}\,\mathrm{Tr}\,(F_{\mu\nu}\,F_{\mu\nu}) + 2\,\mathrm{Tr}\,(j_\mu\,A_\mu). \qquad (6.18)$$

We now turn to a second and probably the most popular definition of a gauge theory as a system possessing a local symmetry. Modification of the fields in a local region of space-time can leave the action unchanged. For electromagnetism this is the usual gauge symmetry under

$$A_\mu \to A_\mu + \partial_\mu \Lambda, \qquad (6.19)$$

where the gauge function Λ is an arbitrary function of the space-time coordinates. In the non-Abelian case, a gauge transformation is specified by a mapping of space into the gauge group. We associate a group element $g(x)$ with each space-time point. In matrix notation, A_μ transforms as

$$A_\mu \to g^{-1} A_\mu g + (\mathrm{i}/g_0)\,g^{-1}\partial_\mu g. \qquad (6.20)$$

To recover the electrodynamic transformation of eq. (6.19), consider $g(x)$

to be a simple phase $\qquad g(x) = e^{-ig_0 \Lambda(x)}$. $\hfill$ (6.21)

Thus we can regard electromagnetism as a $U(1)$ gauge theory. Under eq. (6.20), $F_{\mu\nu}$ transforms particularly simply

$$F_{\mu\nu} \to g^{-1} F_{\mu\nu} g. \tag{6.22}$$

The covariant derivative of eq. (6.13) can be generalized to act on any field transforming under the gauge transformation as some representation of the gauge group. Suppose the field ϕ_i transforms as

$$\phi_i \to R_{ij}(g)\phi_j. \tag{6.23}$$

Here the matrices R_{ij} satisfy the representation property

$$R_{ij}(g) R_{jk}(g') = R_{ik}(gg'). \tag{6.24}$$

For example, the field $F_{\mu\nu}^\alpha$ transforms as the adjoint representation

$$F_{\mu\nu}^\alpha \to R^{\alpha\beta}(g) F_{\mu\nu}^\beta, \tag{6.25}$$

where $\qquad g^{-1}\lambda^\alpha g = R^{\alpha\beta}(g)\lambda^\beta$. $\hfill$ (6.26)

Denote the generating matrices for the representation R by v_{ij}^α such that

$$R_{ij}(e^{i\omega^\alpha \lambda^\alpha}) = (e^{i\omega^\alpha v^\alpha})_{ij}. \tag{6.27}$$

These generators satisfy an analog of eq. (6.6)

$$[v^\alpha, v^\beta] = if^{\alpha\beta\gamma} v^\gamma. \tag{6.28}$$

We now define the covariant derivative of ϕ_i

$$(D_\mu \phi)_i = \partial_\mu \phi_i + ig_0 A_\mu^\alpha v_{ij}^\alpha \phi_j. \tag{6.29}$$

The motivation for this definition is the simple gauge transformation property $\qquad (D_\mu \phi)_i \to R_{ij}(g)(D_\mu \phi)_j$. $\hfill$ (6.30)

Note that for the equation of motion eq. (6.12) to remain simple under a gauge change, we must require that our source transform with the adjoint representation

$$j_\mu^\alpha \to R^{\alpha\beta}(g) j_\mu^\beta \tag{6.31}$$

or, in matrix notation $\qquad j_\mu \to g^{-1} j_\mu g$. $\hfill$ (6.32)

We now turn to a third definition of a gauge theory as a theory of phases. Mandelstam (1962) and Yang (1975) have emphasized that the interaction of a particle with a gauge field involves a phase factor associated with any possible world line that the particle might traverse. In a non-Abelian theory, these path-dependent phase factors become matrices in the gauge group. Whenever a material particle traverses some contour in space-time, its wave function acquires a factor from electromagnetic interactions

$$\psi \to \psi \exp\left(ig_0 \int_P A_\mu \, dx_\mu\right) = U(P)\psi, \tag{6.33}$$

where the integral is along the path in question. This factor is particularly simple for a particle at rest

$$U(P) = \exp(ig_0 A_0 t), \qquad (6.34)$$

where t is the total time length of the path. The particle picks up an extra time oscillation at a rate proportional to its charge and the scalar potential. Thus its energy is increased by the scalar potential. Equation (6.33) generalizes this concept to any Lorentz frame.

For a non-Abelian theory we associate an element of the gauge group with any path. Consider a path

$$x_\mu(s), \quad s \in [0, 1], \qquad (6.35)$$

where s represents some parametrization of the points along the path in question. We define a group element for the portion of the path from $x_\mu(0)$ to $x_\mu(s)$ via the differential equation

$$\frac{d}{ds} U(s) = \frac{dx_\mu}{ds} ig_0 A_\mu U(s). \qquad (6.36)$$

For an initial condition we take

$$U(0) = 1. \qquad (6.37)$$

In eq. (6.36) A_μ is a matrix in the sense of eq. (6.15). We can formally solve this system of equations

$$U(s) = \text{P.O.} \left(\exp\left(ig_0 \int_0^s ds \frac{dx_\mu}{ds} A_\mu \right) \right), \qquad (6.38)$$

where P.O. represents a 'path-ordering' instruction for the non-commuting matrices A_μ. In a power series expansion of the exponential, the matrices are to be ordered as encountered along the path, the largest values of the parameter s being to the left. That the matrix $U(s)$ remains in the gauge group follows because it is a product of group elements associated with infinitesimal pieces of the contour.

Under the local gauge transformation of eq. (6.20), this path-ordered exponential is only sensitive to the gauge function at the endpoints of the path

$$U(s) \rightarrow g^{-1}(x_\mu(s)) U(s) g(x_\mu(0)). \qquad (6.39)$$

Consider the case where the path is a closed contour C. The trace of the group element corresponding to such a contour

$$W(C) = \text{Tr}(U(C)) \qquad (6.40)$$

is independent of the starting point on the contour and is invariant under gauge changes. This is the Wilson loop operator and plays a key role in later chapters. The trace in this definition can be replaced by the character in any representation of the gauge group; however, unless otherwise specified, we use the fundamental defining representation.

We end this chapter with a brief mention of yet another definition of a gauge theory. In a canonical Hamiltonian formalism one would like to write particle interactions in terms of operators involving local fields. Furthermore, discussions of Lorentz invariance are facilitated if these fields transform homogeneously under change of Lorentz frame. A gauge theory is one for which this is impossible (Weinberg, 1965). The interaction Hamiltonian necessarily involves the vector potential A_μ. A Lorentz transformation will in general change the gauge, in which case A_μ transforms inhomogeneously. Covariant gauges such as $\partial_\mu A_\mu = 0$ circumvent this problem but only at the expense of an indefinite metric quantum mechanical space.

That a description with local interactions requires the introduction of potentials is made clear in the Aharonov–Bohm (1959) experiment. A further consequence is the peculiar counting of degrees of freedom with a gauge particle. The potential A_μ in electrodynamics has four components, yet the photon has only two physical polarizations. The longitudinal component is unphysical in that its value depends on gauge choice. The second extra degree of freedom disappears because the time component A_0 is not dynamical. None of the equations of motion involve the time derivative of A_0 and thus its value is a function of the other variables. Elimination of A_0, however, generally introduces non-local objects. Indeed, Mandelstam (1962) has presented a non-local formulation of gauge theory without using potentials, but using the path-ordered integrals discussed above.

A lattice formulation rather severely mutilates Lorentz invariance at the outset. Thus this final definition of a gauge theory is not particularly useful here. The existence of unphysical degrees of freedom does persist on the lattice. We will return to this counting when we discuss the Hamiltonian formulation of lattice gauge theory.

Problems

1. Show that the structure constants $f^{\alpha\beta\gamma}$ defined in equation (6.6) are totally antisymmetric.

2. Verify the gauge transformation property of equation (6.39).

3. What are the generators v^α for the adjoint representation defined in eq. (6.26)?

4. Calculate a rectangular Wilson loop for the field theory of free photons. Using any convenient regulator, show how the leading divergence scales with the loop perimeter. Show that the ratio of two such loops with the same perimeter and number of corners is finite as the cutoff is removed.

7
Lattice gauge theory

In this chapter we introduce Wilson's (1974) elegant formulation of gauge fields on a space-time lattice. The idea is heavily motivated by the concept of a gauge field as a path-dependent phase factor. The basic degrees of freedom are group elements associated with bonds or straight-line paths connecting nearest neighbor pairs of lattice sites. The group element associated with an arbitrary path connecting a sequence of neighboring sites is the group product of these fundamental variables. This particular formulation is also remarkable in that the gauge freedom remains as an exact local symmetry.

Considering a general gauge group G, we associate an independent element of G with each nearest-neighbor pair of lattice sites (i,j)

$$U_{ij} \in G. \qquad (7.1)$$

The indices i and j label the lattice sites at the ends of the bond on which U_{ij} is defined. We suppress those indices associated with the fact that U_{ij} is itself a matrix in the gauge group. On traversing a link in the opposite direction, one should obtain the inverse element

$$U_{ji} = (U_{ij})^{-1}. \qquad (7.2)$$

We can define a vector potential by writing

$$U_{ij} = e^{ig_0 A_\mu a}. \qquad (7.3)$$

Here a is the lattice spacing and the Lorentz index μ is the direction of the given bond. We use the matrix notation for A_μ, which is an element of the Lie algebra of the gauge group. The spatial coordinate x_μ associated with A_μ should be in the vicinity of the link in question; for convenience we take it to lie half way along the bond

$$x_\mu = \tfrac{1}{2}a(i_\mu + j_\mu). \qquad (7.4)$$

In the continuum limit, this choice and the fact that U_{ij} should be path-ordered along the bond become irrelevant conventions.

We need an action to determine the dynamics of these field variables. The Lagrangian should reduce in the continuum limit to the classical Yang–Mills theory of the last chapter. The field strength is a generalized

34

curl of the potential. This suggests using integrals of A_μ around small closed contours. Thus motivated, Wilson proposed that the action should be a sum over all elementary squares of the lattice

$$S = \sum_{\square} S_{\square}. \qquad (7.5)$$

The action on each of these squares or 'plaquettes' is the trace of the product of the group elements surrounding the plaquette

$$S_{\square} = \beta[1 - (1/n) \operatorname{Re} \operatorname{Tr}(U_{ij} U_{jk} U_{kl} U_{li})]. \qquad (7.6)$$

Here the sites i, j, k and l circulate about the square in question. The factor $-1/n$ establishes normalization and sign conventions; n is the dimension of the group matrices. The normalization factor β will be defined later. The additive constant in eq. (7.6) is chosen to make the action vanish whenever the group elements approach the identity. The trace in this equation can be in any representation; for now we only consider the fundamental one.

The demonstration that this action reduces to the usual Yang–Mills theory begins with eq. (7.3) for the U_{ij} in terms of the vector potential. Consider, for example, a plaquette with center at x_μ and oriented in the $(\mu\nu) = (1, 2)$ plane. Writing out eq. (7.6) gives

$$\begin{aligned} S_{\square} = \beta(1 - (1/n) \operatorname{Re} \operatorname{Tr}(\exp &ig_0 A_1(x_\mu - \tfrac{1}{2}a\delta_{\mu 2}) \qquad (7.7)\\ &\times \exp ig_0 A_2(x_\mu + \tfrac{1}{2}a\delta_{\mu 1})\\ &\times \exp -ig_0 A_1(x_\mu + \tfrac{1}{2}a\delta_{\mu 2})\\ &\times \exp -ig_0 A_2(x_\mu - \tfrac{1}{2}a\delta_{\mu 1}))). \end{aligned}$$

We now consider vector potentials smooth enough that we can Taylor expand about x. A little suppressed algebra, which the reader should carry out for himself, yields

$$S_{\square} = \beta(1 - (1/n) \operatorname{Re} \operatorname{Tr} \exp(ig_0 a^2 F_{12} + O(a^4))). \qquad (7.8)$$

Here F_{12} is the field strength tensor including the non-linear terms in A arising from manipulation of the orderings of the exponentials in eq. (7.7). Expanding the exponential, we find

$$S_{\square} = (\beta g_0^2/(2n)) a^4 \operatorname{Tr}(F_{12}^2) + O(a^6). \qquad (7.9)$$

The term of order a^2 vanishes because for unitary groups, the only type considered here, the group generators are Hermitian. We now approximate the sum over all plaquettes with a space-time integral to obtain

$$S = (\beta g_0^2/(2n)) \int \tfrac{1}{2} \operatorname{Tr}(F_{\mu\nu} F_{\mu\nu}) \, d^4x + O(a^6). \qquad (7.10)$$

The factor of one-half under the integral comes from the symmetry under

$\mu\nu$ interchange. Thus we obtain the usual gauge theory action if we identify

$$\beta = 2n/g_0^2. \tag{7.11}$$

The terms with higher powers of the cutoff in eq. (7.10) vanish in the classical continuum limit. Because of divergences in the quantum theory, they can give rise to a finite renormalization of the coupling constant.

We now have our variables and Lagrangian. To proceed to the quantum theory, we insert the action into a path integral

$$Z = \int (\mathrm{d}U)\,\mathrm{e}^{-S(U)}. \tag{7.12}$$

Here we integrate over all possible values for the gauge variables. As they are elements of a compact group, it is natural to use the invariant group measure for this integration. The next chapter discusses this measure in some detail.

Eq. (7.12) defines the partition function for the statistical system motivating this book. Correlation functions are expectation values as discussed in earlier chapters. If H is some function of the field variables U, then its expectation is defined

$$\langle H \rangle = Z^{-1} \int (\mathrm{d}U)\,H(U)\,\mathrm{e}^{-S(U)}. \tag{7.13}$$

In the quantum mechanical Hilbert space, this is the vacuum expectation value of the corresponding time-ordered operator.

Note that we have not included any gauge fixing terms in the path integral. In usual continuum formulations, such terms eliminate a divergence from integrating over all gauges. Here, however, the variables are elements of a compact group. As a consequence, the gauge orbits are themselves compact. For gauge-invariant observables, it is harmless to include an integral over all gauges. We will, however, need to introduce the concept of gauge fixing in order to formulate perturbation theory or to use the transfer matrix to find a Hamiltonian formalism. We will discuss these points later.

Up to this point we have been considering only pure gauge fields. Inclusion of quark degrees of freedom simply involves taking the fermionic action from chapter 3 and inserting a factor of U_{ij} on the fermi field whenever a quark hops from site i to site j. The quark fields have an additional supressed internal symmetry index upon which this matrix acts. Adopting Wilson's projection operator technique for dealing with species doubling, we take eq. (4.55) and write the action for the full interacting gauge theory of quarks and gluons on a lattice

$$S = \beta \sum_{\square} (1 - (1/n) \operatorname{Re} \operatorname{Tr} U_\square)$$

$$+ \tfrac{1}{2} i a^3 \sum_{\{i,j\}} \bar{\psi}_i (1 + \gamma_\mu e_\mu) U_{ij} \psi_j$$

$$+ (a^4 m_0 + 4a^3) \sum_i \bar{\psi}_i \psi_i. \qquad (7.14)$$

Here we have used several shorthand notations to keep the expression manageable. First, $U_\square$ represents the product of group elements around the plaquette in question. Second, the sum over $\{i, j\}$ is over all nearest-neighbor pairs and includes one term for each ordering of i and j. Finally, e_μ represents a unit vector in the direction from site i to site j. We have placed the subscript 0 on the mass m_0 to emphasize that this is the bare mass and will need to be renormalized for a continuum limit of this interacting theory. It is straightforward to introduce other matter fields, such as scalars. As these do not seem to play any role in strong interaction physics, we will only briefly mention them in chapter 9, where we point out some peculiarities of the Higgs mechanism for generating gauge meson masses.

On a kinematic level, the lattice theory has by construction the appropriate classical continuum limit as the Yang–Mills theory. Before such a limit, however, the model still keeps many other aspects of a gauge theory. For one, we work directly with a theory of phases. Furthermore, a local gauge symmetry remains exact. If we associate an arbitrary group element g_i with each lattice site, then the action is invariant under the change

$$U_{ij} \rightarrow g_i U_{ij} g_j^{-1}$$

$$\psi_i \rightarrow g_i \psi_i$$

$$\bar{\psi}_i \rightarrow \bar{\psi}_i g_i^{-1}. \qquad (7.15)$$

Only the definition of a gauge theory in terms of the Lorentz properties of the fields appears to be irrelevant to the lattice formulation, which rather severely mutilates space-time symmetries.

Faithfulness to an exact gauge symmetry should not be a requirement of a cutoff scheme. Indeed, the physics of a renormalizable theory should not depend on the details of the regulator. Nevertheless, this elegant formalism introduced by Wilson greatly simplifies strong coupling treatments of confinement and has been nearly universally adopted in lattice treatments.

Problems

1. Explicitly carry out the steps between eqs (7.7) and (7.10).

2. Consider taking the trace in eq. (7.8) in the adjoint rather than the fundamental representation of the group. What happens to eq. (7.11)?

3. Show that the fermionic terms in eq. (7.14) have the correct classical continuum limit.

8

Group integration

Wilson's use of the invariant measure in his definition of lattice gauge theory lends a flair of mathematical elegance to the subject. This measure is essential to the simplicity of the gauge symmetries in the cutoff theory. In this chapter we review some general properties of invariant integrals over compact Lie groups. We will explicitly display the measure for some simple cases and then discuss integrals over polynomials of $SU(n)$ matrices.

To begin, we must have the basic properties of any integral

$$\int dg\, (af(g) + bh(g)) = a \int dg f(g) + b \int dg\, h(g), \qquad (8.1)$$

$$\int dg f(g) > 0 \quad \text{whenever} \quad f(g) > 0 \quad \text{for all} \quad g. \qquad (8.2)$$

Here f and h are arbitrary functions over the group and a and b are arbitrary complex numbers. We now impose the additional constraint that the measure be left-invariant

$$\int dg f(g) = \int dg f(g'g), \qquad (8.3)$$

where g' is an arbitrary fixed element of the group. In an ordinary integral, this corresponds to a shift of the integration variable. As we will only be considering compact groups, we can normalize the measure such that

$$\int dg\, 1 = 1. \qquad (8.4)$$

We will now show that this measure exists and is unique. We do this by first finding an expression for it under the assumption of its existence, and then we will show that this expression works.

To begin, we consider an arbitrary parametrization of the group elements in terms of a set of parameters α_i where the index i runs from one to n, the dimension of the group manifold. We assume that as the parameters α run over some domain D of R^n, the corresponding group element runs once over the group

$$G = \{g(\alpha) \,|\, \alpha \in D\}. \qquad (8.5)$$

The group multiplication is represented by a function $\alpha(\beta, \gamma)$ satisfying

$$g(\alpha(\beta, \gamma)) = g(\beta)g(\gamma), \tag{8.6}$$

where α, β, and γ all reside in D. We now wish to find a weight $J(\alpha)$ such that the group integral is an ordinary n-dimensional integral

$$\int dg f(g) = \int d\alpha_1 \ldots d\alpha_n \, J(\alpha) f(g(\alpha)). \tag{8.7}$$

The integral on the right hand side of this equation is over the domain D. Writing the group invariance property in this notation gives

$$\int d\beta J(\beta) f(g(\beta)) = \int d\beta J(\beta) f(g(\alpha(\gamma, \beta))), \tag{8.8}$$

where γ parametrizes the factor g' in eq. (8.3). We now change variables to $\alpha(\gamma, \beta)$ with the result

$$\int d\beta J(\beta) f(g(\beta)) = \int d\alpha \left\| \frac{\partial \alpha}{\partial \beta} \right\|^{-1} J(\beta) f(g(\alpha)), \tag{8.9}$$

where $\| \partial\alpha/\partial\beta \|$ represents the Jacobian determinant for the change of variables. Since this is true for arbitrary f, we conclude

$$J(\alpha) = \| \partial\alpha/\partial\beta \|^{-1} J(\beta). \tag{8.10}$$

Taking β to the identity, denoted by e, we find

$$J(\gamma) = K \| \partial(\alpha(\beta, \gamma))/\partial\beta \|^{-1}|_{\beta = e}, \tag{8.11}$$

where $K = J(e)$ is a normalization factor, determined in magnitude with eq. (8.4). Thus the group measure is simply a Jacobian factor. It represents the shift of a small standard volume from near the identity to any point in the group.

If an invariant measure exists, eq. (8.11) is an expression for it. We must now show that this formula works. In particular, eq. (8.10) must be true for all β. We need to show that

$$J(\alpha(\beta, \gamma)) = K \| \partial(\alpha(\delta, \alpha(\beta, \gamma)))/\partial\delta \|_{\delta = e}^{-1} \tag{8.12}$$

is equal to

$$\| \partial\alpha(\beta, \gamma)/\partial\beta \| J(\beta) = K \left\| \frac{\partial\alpha(\beta, \gamma)}{\partial\beta} \right\|^{-1} \left\| \frac{\partial\alpha(\delta, \beta)}{\partial\delta} \right\|^{-1}_{\delta - e} \tag{8.13}$$

For this we need associativity, which implies

$$\alpha(\delta, \alpha(\beta, \gamma)) = \alpha(\alpha(\delta, \beta), \gamma). \tag{8.14}$$

Differentiating with respect to δ gives

$$\left\| \frac{\partial\alpha(\delta, \alpha(\beta, \gamma))}{\partial\delta} \right\| = \left\| \frac{\partial\alpha(\rho, \gamma)}{\partial\rho} \right\|_{\rho = \alpha(\delta, \beta)} \left\| \frac{\partial\alpha(\delta, \beta)}{\partial\delta} \right\|. \tag{8.15}$$

Setting δ to the identity gives the desired result.

Barring a singular parametrization of the group, this analysis proves existence and uniqueness of the measure and provides a formal expression for it. We now show that the right- and left-invariant measures are the same. Clearly a modification of the above arguments will produce a measure which is right-invariant

$$\int (dg)_r f(g) = \int (dg)_r f(gg').$$ (8.16)

Suppose we now define

$$\int (dg)' f(g) = \int (dg)_r f(g_0 gg_0^{-1}),$$ (8.17)

where g_0 is some arbitrary fixed element of the group. This new measure satisfies

$$\int (dg)' f(gg_1) = \int (dg)_r f(g_0 gg_0^{-1} g_1)$$

$$= \int (dg)_r f(g_0 gg_0^{-1}) = \int (dg)' f(g),$$ (8.18)

where we have used the right-invariance of $(dg)_r$. Thus $(dg)'$ is also right-invariant. Uniqueness implies $(dg)' = (dg)_r$. But now we can use right-invariance again in eq. (8.17) to obtain

$$\int (dg)_r f(g) = \int (dg)_r f(g_0 gg_0^{-1}) = \int (dg)_r f(g_0 g).$$ (8.19)

We conclude that the right measure is also left-invariant and, by uniqueness, the measures must be equal. Note that we have used compactness in a rather subtle way. If the integration measures cannot be normalized as in eq. (8.3), the various measures discussed here may differ by constant factors.

We note in passing that

$$\int dg f(g^{-1}) = \int dg f(g).$$ (8.20)

This follows because the left hand side defines another invariant measure which, by uniqueness, must equal the right hand side. In lattice gauge theory, the directions of the bonds do not enter in the measure.

Knowing of its existence may not be useful if the group combination law is complicated. A somewhat more explicit formula for the measure for groups of matrices follows from the definition of a metric tensor on the group

$$M_{ij} = \text{Tr}\,(g^{-1}(\partial_i g)\,g^{-1}(\partial_j g)),$$ (8.21)

where the derivatives are with respect to the parameters α_i

$$\partial_i g = (\partial/\partial\alpha_i)\,g(\alpha).$$ (8.22)

In terms of this metric, the invariant measure is

$$\int \mathrm{d}g f(g) = K \int \mathrm{d}\alpha \,|\det(M)|^{\frac{1}{2}} f(g(\alpha)),\tag{8.23}$$

where the factor of K is again a normalization. This is a standard formula of differential geometry.

We now give some simple examples. For a discrete group the measure is an ordinary sum over the elements. For the group $U(1)$ of relevance to electrodynamics

$$U(1) = \{\mathrm{e}^{\mathrm{i}\theta} \,|\, -\pi < \theta \leqslant \pi\}\tag{8.24}$$

the measure is

$$\int \mathrm{d}g f(g) = \frac{1}{2\pi} \int_{-\pi}^{\pi} \mathrm{d}\theta f(\mathrm{e}^{\mathrm{i}\theta}).\tag{8.25}$$

Functions over the group are periodic functions of the angle θ. Group-invariance is under shifts of phase.

For $SU(2)$ we can parametrize the elements as the surface of a four-dimensional sphere (S_3)

$$SU(2) = \{a_0 + \mathrm{i}\mathbf{a} \cdot \boldsymbol{\sigma} \,|\, a_0^2 + \mathbf{a}^2 = 1\}.\tag{8.26}$$

The matrices $\boldsymbol{\sigma}$ are the Pauli matrices used in chapter 5 when we discussed fermions. With this parametrization the group measure assumes a particularly simple form

$$\int \mathrm{d}g f(g) = \pi^{-2} \int \mathrm{d}^4 a \,\delta(a^2 - 1) f(g).\tag{8.27}$$

Here we use the shorthand notation

$$a^2 = a_0^2 + \mathbf{a} \cdot \mathbf{a}.\tag{8.28}$$

For $SU(3)$ we refer the reader to the discussion by Beg and Ruegg (1965).

For many purposes an explicit form for the measure is unnecessary. In Monte Carlo simulations, to be discussed later, certain algorithms move randomly around in the group in a uniform manner and automatically generate the correct measure. For analytic work, many integrals can often be done using symmetry arguments. For example, the expression

$$\int \mathrm{d}g \, R_{\alpha\beta}(g)\tag{8.29}$$

will vanish if $R_{\alpha\beta}$ is a non-trivial irreducible matrix representation of the group. A group integral selects the singlet part of any function over the group. In particular, we have the relation

$$\int \mathrm{d}g \, \chi_{R_1}(g) \cdots \chi_{R_k}(g) = n_\mathrm{s}(R_1 \otimes \cdots \otimes R_k),\tag{8.30}$$

where the character $\chi_R(g)$ denotes the trace of the matrix corresponding to g in representation R, and $n_\mathrm{s}(R_1 \ldots R_k)$ is the number of times the singlet representation occurs in the direct product of the representations R_1 to R_k. If R and R' are irreducible, we have the orthogonality of the characters

$$\int dg\, \chi_R^*(g)\, \chi_{R'}(g) = \delta_{R,\,R'}. \tag{8.31}$$

For $SU(3)$ we have the integral

$$\int dg\, (\chi_3(g))^3 = 1. \tag{8.32}$$

For the strong coupling expansion we will need integrals of polynomials of the group elements in the fundamental representation. We now turn to a set of graphical rules for the evaluation of such integrals with the groups $SU(n)$ (Creutz, 1978*b*). We are interested in expressions of the form

$$I = \int dg\, g_{i_1 j_1} \cdots g_{i_n j_n} g_{k_1 l_1}^{-1} \cdots g_{k_m l_m}^{-1}, \tag{8.33}$$

where we explicitly indicate the matrix indices on the group elements. It is useful to introduce a generating function for these integrals

$$W(J, K) = \int dg \exp\left(\mathrm{Tr}\,(Jg + Kg^{-1})\right). \tag{8.34}$$

Here J and K are arbitrary n-by-n matrices. To obtain the integral in eq. (8.33), we take derivatives of this generating function

$$I = \left(\frac{\partial}{\partial J_{j_1 i_1}} \cdots \frac{\partial}{\partial K_{l_m k_m}}\right) W(J, K)\big|_{J = K = 0}. \tag{8.35}$$

Invariance of the group measure gives W the symmetry properties

$$W(J, K) = W(K, J) = W(g_0^{-1} J g_1, g_1^{-1} K g_0), \tag{8.36}$$

where g_0 and g_1 are arbitrary $SU(n)$ matrices.

The generating function satisfies an interesting system of differential equations. Since $gg^{-1} = 1$, we have

$$(\partial/\partial K_{ik})(\partial/\partial J_{kj}) W(J, K) = \delta_{ij}. \tag{8.37}$$

And since the determinant of an $SU(n)$ matrix is unity, we have

$$\det(\partial/\partial J)\, W(J, K) = 1. \tag{8.38}$$

Along with the initial condition

$$W(0, 0) = 1, \tag{8.39}$$

these differential equations are sufficient to determine W. Several authors have studied these equations in the large n limit (Brower and Nauenberg, 1980; Bars, 1981). We will solve them iteratively in powers of J and K and give a graphical algorithm for evaluating the coefficients in this expansion.

We first eliminate the K dependence in W using the expression for g^{-1} in terms of the cofactors of g

$$(g^{-1})_{ij} = (\text{cof}(g))_{ij}$$
$$= (1/(n-1)!)\,\epsilon_{j,\,i_1\ldots,\,i_{n-1}}\,\epsilon_{i,\,j_1\ldots,\,j_{n-1}}\,g_{i_1 j_1}\cdots g_{i_{n-1}j_{n-1}}, \qquad (8.40)$$

where ϵ denotes the totally antisymmetric tensor with $\epsilon_{1\ldots n}=1$. This allows us to solve eq. (8.37), replacing derivatives with respect to K by derivatives with respect to J

$$W(J, K) = \exp\left(\text{Tr}\,(K\,\text{cof}\,(\partial/\partial J))\right)W(J), \qquad (8.41)$$

where
$$W(J) = W(J, K = 0) = \int dg\,\exp\left(\text{Tr}\,(Jg)\right). \qquad (8.42)$$

To evaluate $W(J)$ we use the invariance of eq. (8.36), which now reads

$$W(J) = W(g_0^{-1}Jg_1). \qquad (8.43)$$

In an appendix of Creutz (1978*a*) it is proven that any analytic function of J satisfying this symmetry property is a function only of the determinant of J. Thus we expand

$$W(J) = \sum_{i=0}^{\infty} a_i\,(\det J)^i. \qquad (8.44)$$

Normalization of the integration measure implies

$$a_0 = 1. \qquad (8.45)$$

A recursion relation determining further a_i follows from the second differential equation, eq. (8.38). A tedious combinatoric exercise (Creutz, 1978*b*) shows

$$(\det(\partial/\partial J))(\det J)^i = \frac{(i+n-1)!}{(i-1)!}(\det J)^{i-1}. \qquad (8.46)$$

From eqs (8.38), (8.44) and (8.46) we find

$$a_i = \frac{(i-1)!}{(i+n-1)!}a_{i-1}. \qquad (8.47)$$

With eq. (8.45), this is solved to give

$$a_i = \frac{2!3!\ldots(n-1)!}{i!(i+1)!\ldots(i+n-1)!}. \qquad (8.48)$$

Our final power series expression for $W(J)$ is

$$W(J) = \sum_{i=0}^{\infty} \frac{2!\ldots(n-1)!}{i!\ldots(i+n-1)!}(\det J)^i. \qquad (8.49)$$

Note that the determinant of a matrix is simply expressed in terms of the antisymmetric tensor ϵ

$$\det J = (1/n!)\epsilon_{i_1\ldots i_n}\,\epsilon_{j_1\ldots j_n}\,J_{i_1 j_1}\ldots J_{i_n j_n}. \qquad (8.50)$$

A graphical notation is useful for carrying out the derivatives in eq.

(8.35). We use directed line segments to denote group elements. In figure 8.1 we illustrate the convention of upward directed lines representing factors of g while downward lines represent g^{-1}. The ends of these line segments carry as labels the matrix indices of the respective elements. The line direction runs from the first to the second index, as shown in the figure. In figure 8.2 we show how the generic integral from eq. (8.33) appears in this notation.

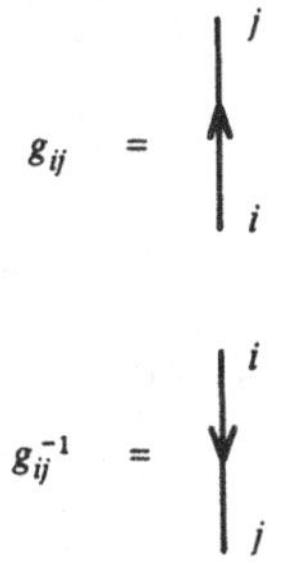

$$g_{ij} \quad = \quad \uparrow\,{}^{j}_{i}$$

$$g_{ij}^{-1} \quad = \quad \downarrow\,{}^{i}_{j}$$

Fig. 8.1. Graphical representation of g and g^{-1} (Creutz, 1978*b*).

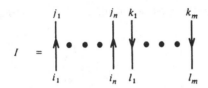

Fig. 8.2. The generic integral under consideration (Creutz, 1978*b*).

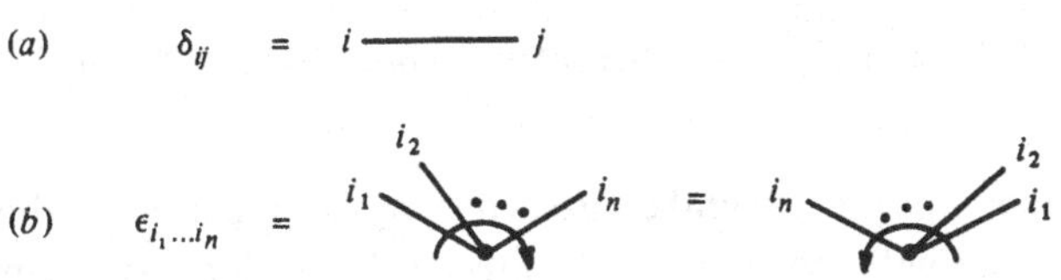

(*a*)　　δ_{ij}　=　i —————— j

(*b*)　　$\epsilon_{i_1 \ldots i_n}$　=

Fig. 8.3. Representation of (*a*) the Kronecker symbol and (*b*) the antisymmetric tensor (Creutz, 1978*b*).

We represent the Kronecker delta symbol δ_{ij} with an undirected line connecting the indices i and j, as shown in figure 8.3*a*. The antisymmetric epsilon symbol $\epsilon_{i_1 \ldots i_n}$ appears as a vertex joining n lines from the indices i_1 to i_n. As the order of these lines is important, we attach to the vertex an arrow running from the first to the last index, as shown in figure 8.3*b*. Finally, whenever two line segments are connected, a matrix product is understood; i.e., the indices associated with the connected ends are summed over.

In the evaluation of group integrals, products of ϵ symbols often occur. Some useful identities involving such products are:

$$\epsilon_{i_1 \ldots i_n} \epsilon_{i_1 \ldots i_n} = n!, \tag{8.51}$$

$$\epsilon_{i, i_1 \ldots i_{n-1}} \epsilon_{j, i_1 \ldots i_{n-1}} = (n-1)! \, \delta_{ij}, \tag{8.52}$$

$$\epsilon_{i, j, i_1 \ldots i_{n-2}} \epsilon_{k, l, i_1 \ldots i_{n-2}} = (n-2)! \, (\delta_{ik} \delta_{jl} - \delta_{il} \delta_{jk}). \tag{8.53}$$

In our graphical notation these relations appear in figure 8.4.

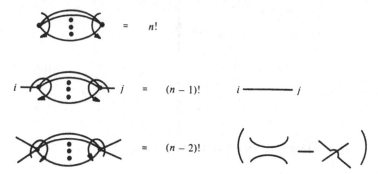

Fig. 8.4. Some combinatoric identities (Creutz, 1978*b*).

Fig. 8.5. Replacing g^{-1} with the cofactors of g (Creutz, 1978*b*).

Evaluation of a group integral consists of a replacement of the directed lines in figure 8.2 with vertices and undirected lines, thus expressing the result in terms of antisymmetric ϵ and Kronecker δ symbols. The first step in this procedure is to convert all directed lines into a set of lines directed only upward. This is accomplished using eq. (8.40), which is shown graphically in figure 8.5. If there were initially more downward than upward lines, it would be simplest to first use eq. (8.20), which says that the arrows on all lines can be simultaneously reversed. Once all lines have the same orientation, we use eqs (8.49) and (8.50) to reduce the integral to a sum of terms involving antisymmetric tensors. Note that the integral automatically vanishes unless the number of group lines is a multiple of n. Supposing we have np lines, where p is an integer, we display eq. (8.49) graphically in figure 8.6. The indicated sum over permutations is over all topologically distinct ways of connecting the group indices to pairs of vertices. The factor in the figure already includes permutations of indices coupled to the same vertex pair and permutations of the vertex pairs. The resulting sum has $(np)!/(p!(n!)^p)$ terms.

Certain identities on the group elements have a simple graphical representation. For example, invariance of the Kronecker symbol

$$g_{ij}\delta_{jk}(g^{-1})_{kl} = \delta_{il},\tag{8.54}$$

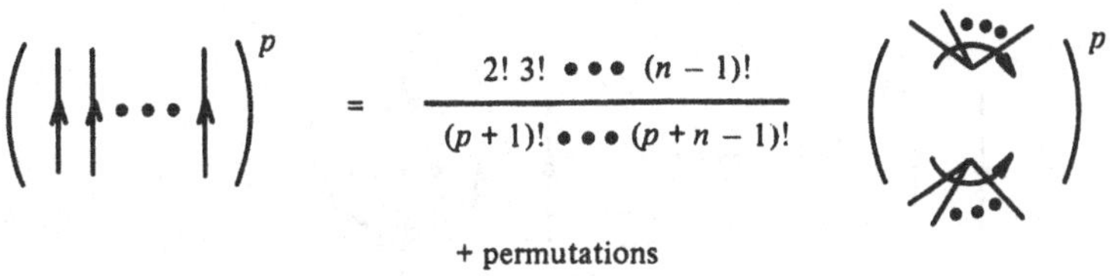

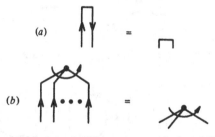

+ permutations

Fig. 8.6. Evaluation of the integral (Creutz, 1978*b*).

Fig. 8.7. Invariance of the (*a*) Kronecker symbol and (*b*) antisymmetric tensor (Creutz, 1978*b*).

is shown in figure 8.7*a*. In terms of the sources J and K, this figure corresponds to eq. (8.37). Invariance of the antisymmetric symbol

$$g_{i_1j_1}\cdots g_{i_nj_n}\epsilon_{j_1\ldots j_n} = \epsilon_{i_1\ldots i_n},\tag{8.55}$$

is shown in figure 8.7*b*. Contracting the indices with an additional ϵ symbol gives the graphical representation of eq. (8.38). Both the identities represented in figure 8.7 are valid regardless of any other lines present in the diagram.

We conclude this chapter with some simple examples to illustrate these rules. First consider $p = 1$ in figure 8.6. This immediately gives

$$\int dg\, g_{i_1j_1}\cdots g_{i_nj_n} = (1/n!)\,\epsilon_{i_1\ldots i_n}\epsilon_{j_1\ldots j_n}.\tag{8.56}$$

In low-order strong coupling expansions a useful integral is

$$I_{ijkl} = \int dg\, g_{ij}(g^{-1})_{kl}.\tag{8.57}$$

This is evaluated graphically in figure 8.8 Here we first use figure 8.5 to
direct all lines upwards, then we use figure 8.6 to eliminate these lines, and
finally we use the identity from figure 8.4 to obtain the result

$$I_{ijkl} = (1/n)\,\delta_{jk}\,\delta_{il}. \qquad (8.58)$$

As a final example consider

$$I = \int \mathrm{d}g \, g_{ij}\,(g^{-1})_{kl}\,g_{mn}\,(g^{-1})_{pq}. \qquad (8.59)$$

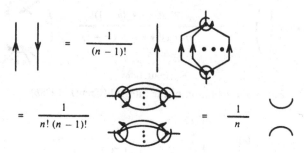

Fig. 8.8. Evaluation of the integral $\int \mathrm{d}g\, g_{ij}\, g_{kl}^{-1}$ (Creutz, 1978*b*).

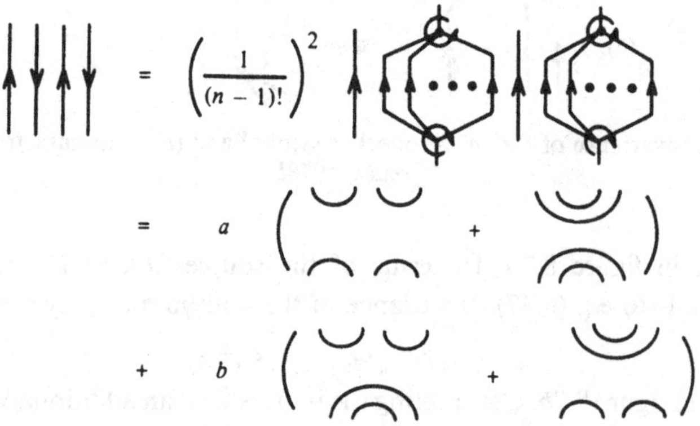

Fig. 8.9. The integral $\int \mathrm{d}g\, g_{ij}\, g_{kl}^{-1}\, g_{mn}\, g_{pq}^{-1}$ (Creutz, 1978*b*).

In figure 8.9 we use figure 8.5 to express I in terms of $2n$ upward lines.
Use of figure 8.6 at this point would give an expression with $(2n)!/(2!n!^2)$
terms. Some simple tricks allow us to simplify this expression for general
n. All terms in this sum have four, an even number, of ϵ vertices both at
the top and at the bottom of the diagram. These can all be eliminated using
identities similar to those in figure 8.4. Thus the result must finally appear
in terms of sets of Kronecker δ symbols connecting separately indices at
the top and bottom of the diagram. Furthermore, note that a Kronecker

δ cannot connect the indices i and m or j and n because they can be initially coupled only through an odd number of ϵ vertices. Thus the final expression for the integral must take the form

$$I = a(\delta_{il}\delta_{mq}\delta_{jk}\delta_{np} + \delta_{iq}\delta_{ml}\delta_{jp}\delta_{nk})$$
$$+ b(\delta_{il}\delta_{mq}\delta_{jp}\delta_{nk} + \delta_{iq}\delta_{ml}\delta_{jk}\delta_{np}), \quad (8.60)$$

where only two independent coefficients are needed because of the $kl \leftrightarrow pq$ symmetry of the integrand. The coefficients a and b can now be determined by multiplying by δ_{jk} and using figure 8.7a to reduce the integral to that

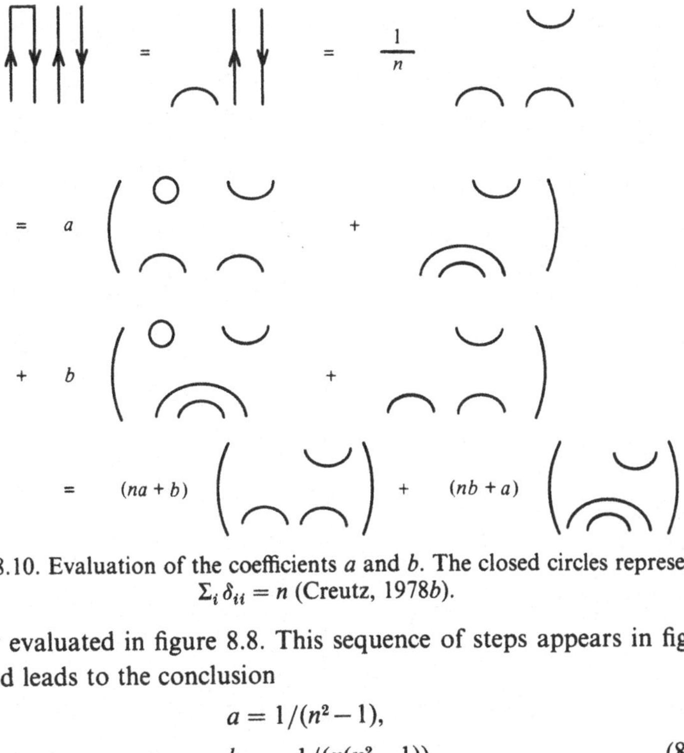

Fig. 8.10. Evaluation of the coefficients a and b. The closed circles represent $\Sigma_i \delta_{ii} = n$ (Creutz, 1978b).

already evaluated in figure 8.8. This sequence of steps appears in figure 8.10 and leads to the conclusion

$$a = 1/(n^2 - 1),$$
$$b = -1/(n(n^2 - 1)). \quad (8.61)$$

Inserted into eq. (8.60), this gives the desired integral.

Problems

1. Show that for 2-by-2 matrices $\det(A) = \frac{1}{2}((\operatorname{Tr} A)^2 - \operatorname{Tr}(A^2))$. What is the corresponding formula for 3-by-3 matrices?

2. For $SU(n)$ evaluate $\int dg \operatorname{Tr}(g^n)$.

3. Show that for irreducible representations R and R'

$$\int dg\, \chi_R^*(g)\chi_{R'}(g_1 g) = d_R^{-1}\delta_{RR'}\chi_R(g_1),$$

where d_R is the dimension of the matrices in the representation.

4. Prove eq. (8.23).

9

Gauge-invariance
and order parameters

For the pure gauge theory without fermions, the formulation of Wilson emphasizes the analogy of lattice gauge theory with models of magnetism in statistical mechanics. The U_{ij} are much like 'spins' located on the bonds of the crystal. These variables then interact through the four-spin coupling in the Wilson action. Further pursuing this analogy, one might ask whether a lattice gauge theory can ever develop a spontaneous magnetization. In a ferromagnet, the spins develop a non-vanishing expectation value in the direction of the magnetization. Thus we might look for phases of lattice gauge theory where

$$\langle U_{ij} \rangle \neq 0. \tag{9.1}$$

We will now show that this is impossible in the Wilson theory.

In an ordinary magnet, such an expectation value represents a spontaneous breaking of a global symmetry. The magnetization has to choose some direction in which to point. This may be determined either with appropriate boundary conditions or with a limit on a vanishingly small applied magnetic field. Once a direction is selected, it remains stable because of the infinite number of degrees of freedom in the thermodynamic limit. Thermal fluctuations cannot coherently shift the magnetization of a large crystal.

In lattice gauge theory, however, an expectation value as indicated in eq. (9.1) breaks the local symmetry of gauge invariance. Because the Wilson action is unchanged under the substitution

$$U_{ij} \rightarrow g_i\, U_{ij}(g_j)^{-1}, \tag{9.2}$$

one can arbitrarily rotate the direction of U_{ij}. As this can be done without changing an infinite number of degrees of freedom, unlike the ferromagnet, thermal fluctuations will induce such rotations and ultimately average over all gauges (Elitzur, 1975). More formally, if we change variables on all other links emanating from site i

$$U_{ik} \rightarrow U_{ij}\, U_{ik}, \quad k \neq j, \tag{9.3}$$

then all dependence on U_{ij} cancels from the action and we have

$$\langle U_{ij} \rangle = \int \mathrm{d}U_{ij}\, U_{ij}, \tag{9.4}$$

which vanishes if U_{ij} contains only non-trivial irreducible representations of the group. The magnetization vanishes in pure lattice gauge theory.

This is unfortunate because in a spin model the magnetization provides a useful order parameter for distinguishing phases. At high temperatures the system is disordered and the magnetization vanishes identically. If at lower temperatures the spins have an expectation value, then we are by definition in a ferromagnetic state. If we can show that at sufficiently low temperatures such a state exists, then we have proven that the system has a phase transition. In lattice gauge theory the expectation of U_{ij} always vanishes and therefore cannot be used to monitor phase changes.

As the problem is intimately entwined with gauge invariance, we should look for a gauge-invariant order parameter. Indeed, as the path integral runs over all gauges, the gauge non-invariant parts of any operator are removed from its expectation value. Thus we will concentrate our attention on quantities which are invariant under eq. (9.2). In the pure gauge theory, the simplest example of such an object is the trace of the product of four links around a plaquette, or essentially the action for the given plaquette. It expectation value represents the internal energy of the corresponding thermodynamic system and is given by a derivative of the partition function

$$P = \langle 1 - (1/n)\operatorname{Tr} U_\square \rangle = \tfrac{1}{6}(\partial/\partial\beta)\log Z. \qquad (9.5)$$

The factor $1/6$ is the ratio of the number of sites to number of plaquettes on a four-dimensional lattice.

The 'average plaquette' P is an order parameter in the sense that it must exhibit singularities of the bulk thermodynamics. However it lacks the useful property of a magnetization in that it never vanishes identically except exactly at zero temperature. We cannot distinguish phases with the average plaquette vanishing in one and not another. Indeed, gauge-invariance precludes any local order parameter from having this property of a magnetization in a spin system. By local we mean involving the expectation of a function of gauge variables in a fixed finite domain of the crystal. Several years before Wilson's work, Wegner (1971) used lattice gauge theory based on the group $Z_2 = \{\pm 1\}$ as an example of a class of models lacking local order parameters and yet having a non-trivial phase structure.

Despite its shortcomings as an order parameter, the average plaquette plays a major role in numerical work where it is the simplest variable to evaluate. Indeed, many transitions are easily seen as jumps or singularities in P as a function of the coupling. For example, in figure 9.1 we show P versus the inverse temperature β for the gauge group Z_2 on a four-dimensional lattice. The points are from Monte Carlo analysis and the curves are based on strong coupling series and duality, all subjects of later discussion. The large jump in P is indicative of the strong first-order phase transition in this model.

A hypothetical unconfined phase of a gauge theory based on a continuous group should contain massless gauge bosons. Using a transfer matrix formalism to determine energies, we define the mass gap as the energy difference between the ground state and the first excited state. This quantity will vanish exactly in an unconfined phase with its free gluons.

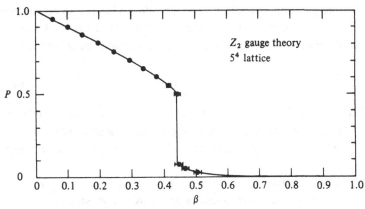

Fig. 9.1. The average plaquette for Z_2 lattice gauge theory. The points are from Monte Carlo simulation and the curves from strong and weak coupling analysis. Note the discontinuity in P at the phase transition at $\beta = \frac{1}{2}\log(1+\sqrt{2})$ (Creutz, 1980a).

In contrast, in a phase displaying confinement of massive quarks, we should have a spectrum of massive glueballs and bound states of quarks. Thus the mass gap is an order parameter which is expected to vanish in one phase but not another. In statistical mechanics language, the mass gap is the inverse of the correlation length. The expectation of two separated operators in a statistical system will generally display a correlation between the operators which falls with the distance between them. If for asymptotic separations this falloff is exponential, then the coefficient of the decrease is the mass gap m

$$C(r) \sim \exp(-mr). \qquad (9.6)$$

This may be justified using a transfer matrix along the separation r. More physically, this equation represents a Yukawa exchange of the lightest particles on the theory. When the mass gap vanishes, we obtain power law forces as familiar in electrodynamics. Note that as an order parameter the mass gap is not local in that its definition involves correlations between asymptotically separated operators.

The use of the mass gap as an order parameter becomes somewhat more complicated if in the confinement phase the hadronic spectrum happens to display a massless particle. This is not simply an academic point because such a behavior is expected when bare quark masses vanish. In this chiral limit, alluded to in chapter 3, γ_5 symmetry is probably manifested in a Nambu–Jona-Lasinio (1961) Goldstone (1961) mode with a vanishing pion mass. In this case a discussion of confinement in terms of the mass gap requires a spin analysis of the massless quanta.

For the pure gluon theory without quarks. Wilson has proposed another non-local order parameter. The trace of a product of links around a closed loop is a gauge-invariant construction. Its expectation value is called the Wilson loop

$$W(C) = \langle \text{Tr} \prod_{ij \in C} U_{ij} \rangle. \qquad (9.7)$$

Here C denotes the loop in question and the group elements are ordered as encountered in circumnavigation of the contour. The simplest non-trivial Wilson loop is the average plaquette, defined in eq. (9.5) with an extra additive constant.

If a quark were to pass around the contour C, its wave function would pick up an internal symmetry rotation given by the product of the link variables encountered. The Wilson loop essentially measures the response of the gauge fields to an external quarklike source passing around its perimeter. For a timelike loop, this represents the production of a quark pair at the earliest time, moving them along the world lines dictated by the sides of the loop, and then annihilating at the latest time. If the loop is a rectangle of dimensions T by R, a transfer matrix argument suggests that for large T

$$W(R, T) \underset{T \to \infty}{\sim} \exp(-E(R)T), \qquad (9.8)$$

where $E(R)$ is the gauge field energy associated with static quark–antiquark sources separated by distance R. If the interquark energy for large separations grows linearly

$$E(R) \underset{R \to \infty}{\to} KR, \qquad (9.9)$$

then we expect for large loops of long rectangular shape

$$W(R, T) \sim \exp(-KRT). \qquad (9.10)$$

The loop expectation falls with the exponential of the area of the loop and the coefficient of this area law is the coefficient of the linear potential. Physically, this area law represents the action of the world sheet of a flux tube connecting the sources. This picture suggests that this area law behavior should hold for arbitrarily shaped loops as long as they are larger than the cross sectional dimensions of a flux tube. In general we expect that with linear confinement

$$W(C) \sim \exp(-KA(C)), \qquad (9.11)$$

where $A(C)$ is the minimal surface area enclosed with the loop C.

In a theory without confinement, the energy of a quark pair should not grow indefinitely with separation, but rather approach twice the self energy of an isolated quark. In such a situation the expectation value of the Wilson loop will decrease more slowly with loop size, in particular exponentially with the perimeter of the contour

$$W(C) \sim \exp(-kp(C)). \qquad (9.12)$$

Here $p(C)$ is the perimeter and k is the self energy contained in the gauge fields around an isolated quarklike source. Some perimeter law behavior should always be present, even in a confining phase where an area law behavior dominates for large enough loops.

The coefficient of the area law provides another order parameter for lattice gauge theory. It vanishes identically in unconfined phases while remaining non-zero whenever quark sources experience a linear long-range potential. It has been extensively studied partly because of its simple flux tube interpretation and partly because of the ease of its evaluation in the strong coupling limit, to be discussed later. As it is directly related to the inter-quark potential, this coefficient is a physically meaningful parameter. In particular, it should be finite in the continuum limit of the pure gluonic theory. This is in contrast with the perimeter law behavior which should contain self energy divergences as the cutoff is removed. The area law is similar to the mass gap in that it represents a non-local order parameter. This is because of its definition in terms of the asymptotic behavior of a correlation function. It has the advantage over the mass gap in that it may be of value even for non-continuous groups such as Z_2 which may lose confinement without the appearance of a massless particle.

The area law criterion for confinement loses its value when quarks are introduced as dynamical variables. In this situation widely separated sources will reduce their energy by creating a pair of quarks from the vacuum fluctuations and screening their long range gauge fields. Effectively, a large Wilson loop measures the potential between two mesons rather than simple bare quarks. If we knew how to calculate with the full theory, however, we would not need a criterion for confinement. All we need to do is calculate the mass spectrum and see if it agrees with laboratory experiments. Hopefully we will soon reach this stage.

Similar interesting questions regarding order parameters arise in gauge theories of the weak interaction, where a Higgs (1964) mechanism generates masses for the gauge bosons. In these theories lattice techniques have played almost no role, primarily because perturbative methods are more than adequate for relevant phenomenology. In the standard presentation, an expectation value for the Higgs field first results in a massless Goldstone (1961) boson which is subsequently 'eaten' by the gauge field and becomes the longitudinal component of a massive vector boson.

On more detailed inspection, this concept of the Higgs field acquiring a vacuum expectation value is overly simplistic. In particular, this field, and thereby its expectation, is not gauge-invariant. In some gauges such as the temporal one the Higgs expectation value is necessarily zero (Creutz and Tudron, 1978; Frohlich, Morchio and Strocchi, 1981) and the vector meson mass is related to the behavior of the vacuum under time-independent gauge transformations which are non-trivial at spatial infinity.

In lattice gauge theory one usually integrates over all gauges. When a Higgs field is present, its direction is thus averaged over. We conclude that the Higgs phase of the theory does not possess a local order parameter in the sense discussed at the beginning of this chapter. As with the confinement question, we could use the mass gap as a non-local order parameter distinguishing the Higgs phase from the massless vector meson phase. But this raises a rather peculiar question. What is the difference between the Higgs and confinement phases? Indeed, both are expected to have mass gaps. Fradkin and Shenker (1979) have shown that in certain cases these phases are not distinct and one can analytically continue from one to the other. This occurs when the Higgs field is in the fundamental representation of the gauge group. In this case the concept of confinement becomes obscured by the fact that an external source can always be screened by Higgs particles. This phenomenon gives rise to an alternative set of words to describe the states in a weak interaction theory when the Higgs fields are in the fundamental representation. For example, the electron would be a confined bound state of a bare electron and a Higgs particle (Abbott and Farhi, 1981a, b).

We now leave the discussion of order parameters and turn to the question of gauge fixing in the lattice theory. In Wilson's formulation, quantization does not require a choice of gauge. The integrals over the link variables are each over a compact domain and thus there cannot be any divergences arising from an integral over all gauges. This contrasts with usual continuum formulations where the volume of the gauge orbits is infinite and some sort of gauge fixing becomes a necessity. In addition to regulating the conventional ultraviolet divergences of field theory, the Wilson prescription also cuts off the total gauge volume. On the other hand, the gauge invariance of the action still permits working within a fixed gauge without affecting the expectations of gauge-invariant operators, such as the Wilson loop. We will now discuss a special class of gauges which are particularly simple in the lattice theory (Creutz, 1977).

Let $P(U)$ be some polynomial in the link variables which is invariant under the general gauge transformation of eq. (9.2). The following discussion goes through unchanged with other fields, such as those of quarks, present; however, for simplicity we consider only the pure gauge theory. Associated with this polynomial is a Green's function

$$G(P) = Z^{-1} \int (dU) e^{-S(U)} P(U). \tag{9.13}$$

We begin the discussion with the consideration of a single link from site i to site j. Suppose that in evaluating the expectation in eq. (9.13) we forgot to integrate over that one link variable. Remarkably, we will now see that the result for $G(P)$ would not be affected by our sloppiness. To see this formally we introduce a delta function on the gauge group. This has the properties

$$\int dg\, \delta(g', g) f(g) = \int dg\, \delta(g, g') f(g) = f(g')$$

$$\delta(g, g') = \delta(g_0 g g_1, g_0 g' g_1) \tag{9.14}$$

for arbitrary g_0 and g_1. Leaving link U_{ij} fixed at the element g rather than integrating over it as instructed in eq. (9.13) gives for the expectation of P

$$I(P, g) = Z^{-1} \int (dU)\, \delta(U_{ij}, g) e^{-S(U)} P(U). \tag{9.15}$$

Clearly if we integrate over g we get back to eq. (9.13)

$$G(P) = \int dg\, I(P, g), \tag{9.16}$$

If we now consider the gauge transformation of eq. (9.2) and note the invariance of $S(U)$, $P(U)$, and the measure, we obtain

$$I(P, g) = I(P, g_i^{-1} g g_j). \tag{9.17}$$

Since g_i and g_j are arbitrary, we conclude that $I(P, g)$ is actually independent of g. Eq. (9.16) then tells us

$$I(P, g) = G(P), \tag{9.18}$$

which is what we set out to prove. To calculate a gauge-invariant Green's function we can set any particular link variable to an arbitrary group element and only integrate over the remaining variables.

The above process can be repeated to fix more link variables. The final result is that we can arbitrarily neglect to integrate over any set of U_{ij} as long as this set contains no closed loops. The fixed links should form a tree, which may be disconnected. The gauge is completely determined if we have a maximal tree, a tree to which the addition of any more links would create a closed loop. An example of such a maximal tree is shown

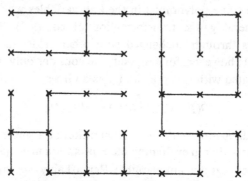

Fig. 9.2. An example of a maximal tree. All links on the tree can be set arbitrarily by the gauge fixing process.

in figure 9.2. The U_{ij} can be set to arbitrary group elements g_{ij}. The general formula for the Green's function of our gauge-invariant operator is

$$G(P) = Z^{-1} \int (\mathrm{d}U) \prod_{\{ij\} \in T} \delta(U_{ij}, g_{ij}) \, \mathrm{e}^{-S(U)} P(U). \qquad (9.19)$$

Here T denotes the tree in question and $\{ij\}$ refers to the link connecting sites i and i with arbitrary orientation.

A particularly simple gauge corresponds to setting all links in a particular direction to unity. This corresponds to an axial gauge where one component of the vector potential vanishes. Choosing the time direction, we obtain the $A_0 = 0$ or temporal gauge. This gauge will be useful for the construction of a transfer matrix and a Hamiltonian formulation of the lattice gauge theory. This gauge is illustrated in figure 9.3 and still leaves the freedom of time-independent gauge transformations.

Note that in an axial gauge plaquettes parallel to that axis represent a simple two-spin coupling of the unfixed variables. The theory reduces to a set of one-dimensional spin chains interacting with each other via the four-spin coupling of the remaining plaquettes. In two space-time dimensions there is no interchain coupling and the pure gauge theory is equivalent to an exactly solvable one-dimensional spin system.

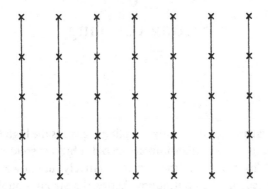

Fig. 9.3. A tree corresponding to the temporal gauge. Here the vertical direction represents time.

Problems

1. Solve two-dimensional lattice gauge theory for pure gauge fields. Find an expression for the average plaquette in terms of simple integrals over the gauge group. Show that the model has no phase transitions. Show that the Wilson loops always exhibit an area law.

2. Consider lattice gauge theory defined by replacing $U_\square$ by the product of links around one-by-two rectangles and with the action being a sum over all such rectangles. Show that the two-dimensional model is no longer trivial. Show that the two-dimensional Z_2 model has a phase transition.

3. Find a gauge fixing tree such that most of the unfixed links have a non-vanishing expectation value, even on an infinite lattice.

4. Given an arbitrary gauge fixing function $f(U)$, show that our gauge-invariant Green's function is given by

$$G(P) = Z^{-1} \int (\mathrm{d}U)(f(U)/\phi(U)) \, \mathrm{e}^{-S} P(U),$$

where the Fadeev–Popov (1967) correction factor $\phi(U)$ is an integral of f over all gauges (Kerler, 1981b)

$$\phi(U) = \int (\prod_i \mathrm{d}g_i) f(g_i \, U_{ij} \, g_j^{-1}).$$

Show that $\phi = 1$ for the gauge fixing function in eq. (9.19).

10
Strong coupling

In the statistical analog, the strong coupling regime is the high temperature limit. High temperature expansions are an old subject in solid state physics, but before Wilson's work they were relatively unknown to particle theorists. Indeed, in the continuum theory the strong coupling limit is rather unnatural and difficult to treat. In contrast, on the lattice strong coupling is by far the simplest limit. One merely expands the Boltzmann factor in powers of the inverse temperature and evaluates the terms in the resulting series. In the gauge theory each power of β is associated with a plaquette somewhere in the lattice. This gives a simple diagrammatic interpretation in terms of graphs built up from such plaquettes (Wilson, 1974; 1975; Balian, Drouffe and Itzykson, 1975b).

We begin our discussion with a rectangular Wilson loop in pure $SU(n)$ lattice gauge theory without fermions

$$W(I, J) = Z^{-1} \int (\mathrm{d}U)\, e^{-S} (1/n)\, \mathrm{Tr} \prod_{ij \in C} U_{ij}. \tag{10.1}$$

Here the curve C is the rectangle of dimensions I by J in lattice units, and the factor of $1/n$ is inserted for convenience. As usual, the group elements are ordered as encountered in circumnavigating the contour. In figure 10.1 we show such a curve for a three-by-three loop. Because the variables become random in the strong coupling limit, it is simplest to shift the action by a constant from the normalization used in eq. (7.6). Thus we take

$$S = -\sum_{\square} (\beta/n) \quad \mathrm{Re\, Tr}\, U_{\square}$$

$$= -\sum_{\square} (\beta/(2n)) (\mathrm{Tr}\, U_{\square} + \mathrm{Tr}\, U_{\square}^*). \tag{10.2}$$

We now observe that because

$$\int (\mathrm{d}U)\, U_{ij} = 0 \tag{10.3}$$

all Wilson loops will vanish as β goes to zero. Indeed, for each link in the contour we must bring down at least one corresponding link from an expansion of the exponential of the action if we are to avoid the zeros from

60

eq. (10.3). Correspondingly, every link from the action must have a partner, either from the action itself or the inserted loop. The first non-vanishing contribution in the strong coupling series comes from tiling the loop with plaquettes as shown in figure 10.2. Note the orientations of the loops in the figure; this is important for all $SU(n)$ except $SU(2)$, for

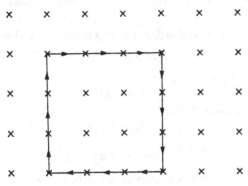

Fig. 10.1. A three-by-three Wilson loop.

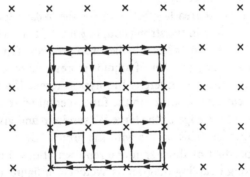

Fig. 10.2. Tiling the loop with plaquettes.

which Tr $U_\square$ is real. The simple integrals needed to evaluate this diagram are

$$\int dg\, 1 = 1 \tag{10.4}$$

for links outside the tiled region and

$$\int dg\, U_{ij}\, U_{kl}^{-1} = (1/n)\, \delta_{il}\, \delta_{jk} \tag{10.5}$$

within the loop. These integrals can all be combined graphically using the rules from the chapter on group integration. We obtain a factor of n^{-1} from each pair of bond variables and a factor of n from each site on the surface, including the boundary. Multiplying these with $\beta/(2n)$ for each plaquette brought down from the exponential of the action, we obtain the result

$$W(I,J) \underset{\beta \to 0}{\to} \begin{cases} (\beta/(2n^2))^{IJ}, & n>2 \\ (\beta/4)^{IJ}, & SU(2). \end{cases} \tag{10.6}$$

The difference for $SU(2)$ arises from the non-oriented nature of the plaquettes.

At this lowest non-trivial order of strong coupling we already see an area law

$$W(I,J) \sim e^{-KA}, \tag{10.7}$$

where the area in physical units is

$$A = a^2 IJ \tag{10.8}$$

and the string tension starts out

$$K \underset{\beta \to 0}{\to} \begin{cases} -a^{-2}\log(\beta/(2n^2)), & n>2 \\ -a^{-2}\log(\beta/4), & SU(2). \end{cases} \tag{10.9}$$

This area law behavior will persist for arbitrarily shaped loops. The leading contribution in the small β limit always follows from tiling the minimal surface bounded by the loop.

The existence of an area law behavior for the Wilson loop is a nearly universal phenomenon in the strong coupling limit. It occurs for all gauge groups in which no singlets appear in the direct product of the fundamental representation with any number of adjoint representations. This includes most but not all groups of interest. In physical terms, if a finite number of gluons can neutralize a source in the fundamental representation, then they will surround the edge of the large Wilson loop and give a perimeter law type of behavior. This occurs with the group $SO(3)$ where a singlet occurs in the product of three spin-one representations. For large loops the leading strong coupling diagram is sketched in figure 10.3. This is a purely gluonic analog of the phenomenon discussed in chapter 9, where we argued that the area law is no longer a useful order parameter after quarks enter the theory.

To keep from writing equations for several cases, we now restrict ourselves to the group $SU(3)$. Then the next contribution arises from replacing one of the tiling plaquettes with two of the opposite orientation. The resulting diagram appears in figure 10.4. The new integrals follow from eq. (8.56) of the chapter on group integration. Allowing for the insertion to be placed anywhere on the tiled surface, we obtain

$$W(I,J) = (\beta/18)^{IJ}(1 + IJ\beta/12 + O(\beta^2)), \tag{10.10}$$

$$a^2 K = -\log(\beta/18) - \beta/12 + O(\beta^2). \tag{10.11}$$

Note that this particular correction to the leading order has the same geometric structure. We still consider the tiling of a minimal surface inside the loop and have only introduced a new type of tile. A simple change of variables permits the summation of all contributions of this type. Consider

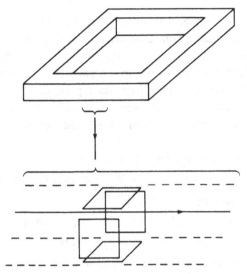

Fig. 10.3. A strong coupling diagram for *SO*(3) gauge theory. This contribution falls exponentially with the perimeter of the loop.

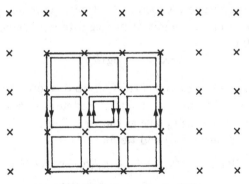

Fig. 10.4. A new type of tile.

the character expansion of the exponentiated plaquette operator

$$\exp\left(\tfrac{1}{3}\beta \operatorname{Re} \operatorname{Tr} U_\square\right) = N(\beta)\left(1 + \sum_{R \neq 1} b_R(\beta)\chi_R(U_\square)\right). \qquad (10.12)$$

Here the sum extends over all non-trivial irreducible representations of the group and χ_R is the trace or character in the corresponding representation. This sum is easily inverted using the orthogonality of the characters (eq. 8.31), with the result

$$N(\beta) = \int dU \exp\left(\tfrac{1}{3}\beta \operatorname{Re} \operatorname{Tr} U\right), \tag{10.13}$$

$$b_R(\beta) = N^{-1} \int dU (\chi_R U)^* \exp\left(\tfrac{1}{3}\beta \operatorname{Re} \operatorname{Tr} U\right). \tag{10.14}$$

Using eq. (10.12), one replaces a sum over arbitrary powers of $\operatorname{Tr} U_\square$ on any given plaquette with a sum over representations of the group, each occurring only once. For low orders in the strong coupling expansion one can rapidly perform the needed group integrals with the use of familiar combination rules to form singlets from the representations appearing on the adjacent plaquettes to any given link. The disadvantage of this method is that as the order increases we must keep track of higher and higher representations.

We will now illustrate this technique with an evaluation of the string tension a^2K to an effective order β^6. For $SU(3)$ the first few coefficients in eq. (10.12) are

$$b_3 = b_{\bar{3}} = \beta/6 + O(\beta^2), \tag{10.15}$$

$$b_6 = b_{\bar{6}} = \beta^2/72 + O(\beta^3) = b_3^2/2 + O(b_3^3), \tag{10.16}$$

$$b_8 = \beta^2/36 + O(\beta^3) = b_3^2 + O(b_3^3). \tag{10.17}$$

Higher representations start with higher powers of β. To avoid needing further terms in eq. (10.15), we express the result in powers of b_3. Note that the normalization $N(\beta)$ drops out of the calculation due to the division by Z in expectation values. As before, the leading term for our flat Wilson loop arises from tiling the minimal surface with fundamental plaquettes. Thus eq. (10.9) becomes

$$a^2K \underset{\beta \to 0}{\to} -\log(b_3/3). \tag{10.18}$$

However now we encounter no corrections to this formula until order b_3^4. This next term comes from a non-minimal tiled surface obtained by placing a cubical bump on our tiled plane as shown in figure 10.5. This adds four new plaquettes to the surface and we find

$$W(I, J) = (b_3/3)^{IJ}(1 + 4IJ(b_3/3)^4 + O(b_3^5)), \tag{10.19}$$

$$a^2K = -\log(b_3/3) + 4(b_3/3)^4 + O(b_3^5). \tag{10.20}$$

The factor of four in front of the new term represents the fact that the bump on our surface can either project above or below the plane in either of the two remaining dimensions of our four-dimensional space-time.

The next contribution arises from the same basic picture as in figure 10.5 but now with a non-trivial representation for the base of the cube. If we put a 'floor' on the bump using the $\bar{3}$ representation and reverse the orientation of all plaquettes in the cap, we find

$$a^2K = -\log{(b_3/3)} + 4(b_3/3)^4 + 12(b_3/3)^5 + O(b_3^6). \qquad (10.21)$$

With the bump's floor in the sextet or octet representations, and with an appropriately oriented cap, we obtain contributions proportional to b_6 or b_8 multiplied by b_3^4. By eq (10.16, 17) these terms are effectively of order b_3^6.

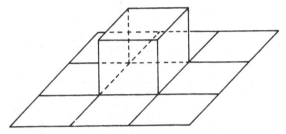

Fig. 10.5. An order b_3^4 correction to the string tension.

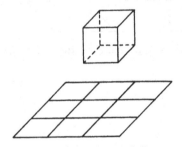

Fig. 10.6. A disconnected diagram.

At order b_3^6 a new type of contribution arises from the division by Z in evaluating expectation values. The partition function itself is a sum over disconnected diagrams. It serves to cancel diagrams with non-trivial representations on clusters of plaquettes completely isolated from the sources in the Wilson loop. For example, the diagram in figure 10.6 need not be evaluated. However, the division also removes some extra pieces if the cluster contributing to Z overlaps the connected numerator diagram. This gives a negative $O(b_3^6)$ contribution to the string tension, as illustrated in figure 10.7.

To complete the sixth-order strong coupling expansion for the string tension, we also must include the non-minimal bump illustrated in figure 10.8. Combining all contributions, we obtain

$$a^2K = -\log{(b_3/3)} + 4(b_3/3)^4 + 12(b_3/3)^5 - 10(b_3/3)^6 + O(b_3^7). \qquad (10.22)$$

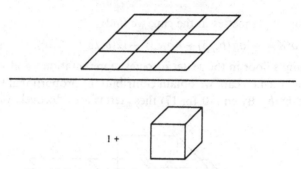

1 +

Fig. 10.7. A contribution from the division by Z.

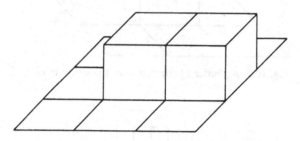

Fig. 10.8. A larger bump on the tiled surface.

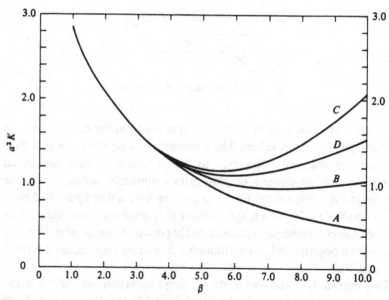

Fig.10.9. The first few strong coupling approximations to the string tension. The curves *A*, *B*, *C*, and *D* respectively correspond to order 3, 4, 5, and 6 in powers of b_3.

Beyond this order the calculation becomes rapidly more tedious. Munster and Weisz (1980) have evaluated the coefficients to order b_3^{12}.

In figure 10.9 we plot the first four strong coupling expressions for $a^2 K$ as functions of the basic inverse charge β. Note that for β less than five the result appears rather stable. This suggests that the radius of convergence of the strong coupling series is of order five. Indeed, the theory is known to be analytic in the vicinity of vanishing β (Osterwalder and Seiler, 1978). This contrasts with the usual perturbative expansion in coupling, which is known to be at best asymptotic (Dyson, 1952; Lipatov, 1977; Brezin, Le Guillou and Zinn-Justin, 1977a,b).

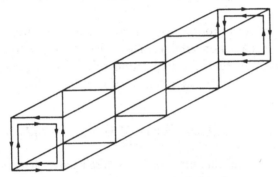

Fig. 10.10. A strong coupling diagram for calculating the mass gap. The sides of the square tube are to be tiled with fundamental plaquettes.

In this purely gluonic theory another interesting quantity for strong coupling studies is the mass gap. This is most easily extracted from the exponential decay of the correlation between two plaquettes separated by a large distance. Thus we effectively study the effects of glueball exchange between two gauge-invariant operators. To leading order we have the diagram shown in figure 10.10, where the tube connecting the end plaquettes is tiled with fundamental plaquettes. This gives

$$am_g = -4\log(b_3/3) + O(b_3^2). \tag{10.23}$$

Using a similar analysis to that for the string tension, Munster (1980) gives the coefficients in this expansion to order b_3^8.

Including the fermions in the strong coupling expansion is a straightforward procedure. We recall the full action from chapter 7

$$S = \beta \sum_{\square} (1 - (1/n) \operatorname{Re} \operatorname{Tr} U_{\square})$$

$$+ \tfrac{1}{2} i a^3 \sum_{\{i,j\}} \bar{\psi}_i (1 + \gamma_\mu e_\mu) U_{ij} \psi_j + (a^4 m_0 + 4a^3) \sum_i \bar{\psi}_i \psi_i. \tag{10.24}$$

In the strong coupling limit the gauge fields become random. This suggests treating the second term in eq. (10.24) as a perturbation. This 'hopping' term represents the movement of quarks between neighboring sites. Thus we begin with the static quark theory defined by the last term in the action and expand the exponential of the action in powers of the remaining terms. Strictly speaking, the resulting perturbation series is not solely in powers of g_0^{-2} because when a quark–antiquark pair moves as a unit from one site to another, the gauge fields can cancel out, regardless of how random they are. Even in the limit of $g_0^{-2} = 0$, the theory is not exactly solvable and we must make a further expansion. The additional perturbative parameter is effectively the inverse quark mass in units of the lattice spacing.

At this point we introduce sources coupled to the various fields, as discussed in the chapter on fermions. This will allow us to reduce the strong coupling expansion to the manipulation of creation and annihilation operators. We add to the action terms linear in the external sources and in the field variables

$$S_I = S + \sum_i (b_i \psi_i - \bar{\psi}_i c_i) + \sum_{\{ij\}} \text{Tr}\,(U_{ij} J_{ij} + U_{ji} \bar{J}_{ij}). \qquad (10.25)$$

Here b and c are anticommuting sources as discussed in chapter 5. For the gauge variables we introduce matrix valued sources J and $\bar{J}$ analogous to the quantities J and K used in the generating function for group integrals in chapter 8. Now, however, there are independent sources for every link in the lattice. As before, we obtain Green's functions from derivatives with respect to the sources. We represent these derivatives as creation operators b^+, c^+, J^+, and $\bar{J}^+$ satisfying the commutation relations

$$[b_i, b_j^+]_+ = \delta_{i,j}, \qquad (10.26)$$

$$[c_i, c_j^+]_+ = \delta_{i,j}, \qquad (10.27)$$

$$[J_{ij}, J_{kl}^+] = \delta_{i,k}\,\delta_{j,l}, \qquad (10.28)$$

$$[\bar{J}_{ij}, \bar{J}_{kl}^+] = \delta_{i,k}\,\delta_{j,l}, \qquad (10.29)$$

with all other commutators or anticommutators, as appropriate, vanishing. We will turn off the sources by applying them to the 'empty vacuum' state satisfying

$$b_i|0\rangle = c_i|0\rangle = J_{ij}|0\rangle = \bar{J}_{ij}|0\rangle = 0. \qquad (10.30)$$

We remind the reader that in eqs (10.26–30) we supress spinor and internal symmetry indices. The site indices indicated explicitly here must not be confused with matrix indices on J and $\bar{J}$.

As in chapter 5, we define a generating state

$$(W| = (0| \int (\mathrm{d}\psi\,\mathrm{d}\bar\psi\,\mathrm{d}U)\exp(-S_I).$$ (10.31)

Green's functions follow from the formula

$$\langle \psi_{i_1}\dots\psi_{i_n}\bar\psi_{j_1}\dots\bar\psi_{j_n}U_{k_1 l_1}\dots U_{k_m l_m}\rangle$$
$$= Z^{-1}(W|b_{i_1}^+\dots b_{i_n}^+ c_{j_1}^+\dots J_{k_m l_m}^+|0),$$ (10.32)

where the partition function is simply

$$Z = (W|0).$$ (10.33)

Because of the forthcoming analogy with the string model, we call the space in which these creation and annihilation operators act 'string space'. The operator J_{ij}^+ creates a string bit pointing from site i to site j. The source b_i^+ creates an antiquark and c_i^+ a quark at site i. Of course one must not confuse these 'quark' states in string space with states in the physical Hilbert space of the Minkowski world. When we discuss the Hamiltonian formulation of lattice gauge theory, states in the latter space will be distinguished by angular brackets, $|\psi\rangle$.

Strong coupling perturbation theory begins with a breakup of the action into two parts
$$S = S_0 + S',$$ (10.34)

where S_0 is the static quark action

$$S_0 = (a^4 m_0 + 4a^3)\sum_i \bar\psi_i\psi_i$$ (10.35)

and S' contains the remaining terms in eq. (10.24). We now write the generating state in the form

$$(W| = (W_0|\exp(-S'(b^+,c^+,J^+,\bar J^+)).$$ (10.36)

Here in S' all dependences on ψ, $\bar\psi$, and U variables have been replaced with the corresponding source-creation operators. The unperturbed generating state is

$$(W_0| = (0| \int (\mathrm{d}\psi\,\mathrm{d}\bar\psi\,\mathrm{d}U)\exp\left(-S_0 - \sum_i(b_i\psi_i - \bar\psi_i c_i)\right.$$
$$\left. - \sum_{\{ij\}}\mathrm{Tr}(U_{ij}J_{ij} + U_{ji}\bar J_{ij}).\right.$$ (10.37)

Since S_0 is quadratic in the anticommuting fields, the fermionic integral is easily done with a completion of the square

$$\int(\mathrm{d}\psi\,\mathrm{d}\bar\psi)\exp(-S_0 - \sum_i(b_i\psi_i - \bar\psi_i c_i)) = N\exp(-\sum_i b_i(4a^3 + ma^4)^{-1}c_i).$$ (10.38)

The irrelevant normalization factor N is just the product over all sites of

$(a^4 m_0 + 4a^3)$. The integral over gauge fields in eq. (10.37) was extensively discussed in chapter 8 for a single link. Thus we have

$$\int (\mathrm{d}U) \exp\left(\sum_{\{ij\}} \mathrm{Tr}\,(U_{ij} J_{ij} + U_{ji} \bar{J}_{ij})\right) = \prod_{ij} W_G(J_{ij}, \bar{J}_{ij}), \qquad (10.39)$$

where W_G is the group-integration generating function of eq. (8.41).

Putting all factors together, we obtain the expression for the generating state

$$\begin{aligned}
(W| = (0|\exp\left(-\sum_i b_i (4a^3 + ma^4)^{-1} c_i\right) \prod_{ij} W_G(J_{ij}, \bar{J}_{ij}) \\
\times \exp\left(-\tfrac{1}{2}\mathrm{i}a^3 \sum_{\{ij\}} c_i^+ (1 + \gamma_\mu e_\mu) J_{ij}^+ b_j^+\right) \\
\times \exp\left(-\beta \sum_\square (1 - (1/n)\,\mathrm{Re\,Tr} \prod_{\{ij\}\in\square} J_{ij}^+)\right). \qquad (10.40)
\end{aligned}$$

The strong coupling expansion follows from a power series treatment of the last two terms.

The four terms in eq. (10.40) have a simple interpretation in string space. The first term destroys quark–antiquark pairs at a single site, the second term destroys sets of string bits associated with nearest-neighbor pairs of sites, the third term creates quark–antiquark pairs separated by one lattice spacing and connected by a string bit pointing to the antiquark, and the last term creates elementary squares of string bits. This creation and destruction of quarks and string bits provides the basis of the diagrammatic rules.

Consider some particular Green's function as in eq. (10.32). The graphical rules for calculating this quantity are read off from eq. (10.40):

(1) Draw a set of string bits, quarks, and antiquarks as created by the corresponding operators in eq. (10.32).

(2) Using the third factor in eq. (10.40), create string bits connecting quark–antiquark pairs to produce a configuration where every site has an equal number of quarks and antiquarks. With several types of quarks, each species must balance separately. Closed quark loops can also be generated at this stage. Every quark–string–antiquark combination generated by this rule gives a factor of $\tfrac{1}{2}\mathrm{i}a^3(1 + \gamma_\mu e_\mu)$ to the amplitude. The spinor indices on these gamma matrices will be contracted in rule (4).

(3) Use the last factor or 'plaquette term' in equation (10.40) to create elementary squares of string bits, thus generating a configuration where every nearest neighbor pairs of sites i,j has a set of string bits which can form a singlet in the gauge group. Thus for $SU(3)$ the number of bits from i to j minus the number from j to i must be a multiple of three. Each plaquette gives a factor of $\beta/6$ to the diagram. A set of m identical plaquettes gives an additional factor of $1/m!$. Alternatively, we can use

the parameters b_R of eq. (10.13) and dress the plaquettes in various representations taken one at a time.

(4) The first term in eq. (10.40) now serves to connect the quark and antiquark lines. They are paired up at each site individually, and in the process spinor, flavor, and the internal symmetry indices of the string bits are contracted. Each such 'quark connection' gives a factor of $(4a^3 + ma^4)^{-1}$ to the amplitude.

(5) At this point we have built up the full diagram. We now begin to tear it down by doing the group integrals. For this purpose we may use the graphical rules from the chapter on group integration. If we are using the parameters b_R, then these integrals proceed as in the discussion of the string tension at the beginning of this chapter.

(6) Some factors of minus one arise from the fermionic nature of the quarks. Each quark line forming an internal closed loop gives a factor of -1. With the Green's function in the standard ordering of eq. (10.32), if each ψ_{i_k} is connected by a quark line to $\bar{\psi}_{j_k}$, then there are no more factors; otherwise we must multiply by minus one to the number of transpositions necessary to put the ψ's in the same order as the $\bar{\psi}$'s they are connected to. This is the same rule which gives an ordinary Feynman diagram an extra minus sign for each interchange of external fermion lines.

(7) Sum over all distinct strong coupling diagrams up to the order desired.

(8) Divide by Z, the sum of all vacuum fluctuation diagrams. This will first of all remove contributions of totally disconnected parts of a diagram. In addition, as noted in the discussion of the string tension, non-trivial contributions arise when the vacuum fluctuation overlaps the diagrams in the numerator.

We now illustrate these rules with a simple example. Taking a single quark species, we study

$$\langle \bar{\psi}_i \gamma_5 \psi_i \bar{\psi}_j \psi_5 \psi_j \rangle. \tag{10.41}$$

This is the two-point function for the composite pseudoscalar field $\bar{\psi}_i \gamma_5 \psi_i$. Rule (1) instructs us to place quark–antiquark pairs at site i and site j as illustrated in figure 10.11. In this figure we let the vertical direction represent x_0 and the horizontal direction represent x_1. In figure 10.12 we show one possible way of applying rule (2), thus adding quark–string–antiquark combinations so as to have all quarks paired with antiquarks. One dressing of the diagram with plaquettes following rule (3) is shown in figure 10.13. Making the quark connections with rule (4) gives figure 10.14. Finally rule (5) is carried out with the repeated use of figure 8.8 to give figure 10.15. Combining the various factors, we obtain the contribution

Fig. 10.11. The quark–antiquark pairs created by the correlation function in eq. (10.41) (Creutz, 1978a).

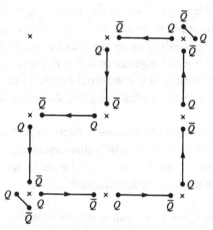

Fig. 10.12. A set of quark–string–antiquark combinations pairing all quarks with antiquarks (Creutz, 1978a).

of this diagram to the amplitude

$$(ia^3)^8 \left(\frac{\beta}{6}\right)^3 (4a^3 + ma^4)^{-10} 3^{-2} \operatorname{Tr} \Gamma, \qquad (10.42)$$

where Γ is the product of the Dirac operators around the diagram

$$\operatorname{Tr} \Gamma = 2^{-8} \operatorname{Tr} (\gamma_5 (1 + \gamma_0)(1 + \gamma_1)(1 + \gamma_0)(1 + \gamma_1),$$

$$\gamma_5 (1 - \gamma_0)^2 (1 - \gamma_1)^2) = \tfrac{1}{4}. \qquad (10.43)$$

Note that eq. (10.42) can be put in the form

$$3(4a^3 + ma^4)^{-2} \operatorname{Tr} \Gamma (4 + ma)^{-p} (\beta/18)^4, \qquad (10.44)$$

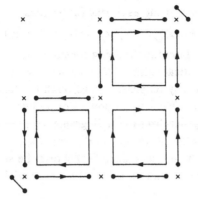

Fig. 10.13. Dressing the diagram with plaquettes (Creutz, 1978*a*).

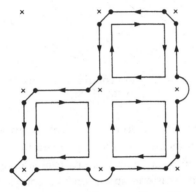

Fig. 10.14. Making the quark connections (Creutz, 1978*a*).

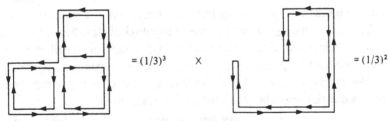

Fig. 10.15. Evaluation of the group integrals (Creutz, 1978*a*).

where p is the total quark line length in units of the lattice spacing and A is the area of the surface covered by plaquettes, measured in units of a^2. This form provides the basis for the string analogy discussed below. Note also the high degree of cancellation of the factors of lattice spacing in eq. (10.42). Because of this the fermionic fields are often rescaled to give a factor of unity with each quark connection of rule (4). Then whenever a quark 'hops' from site to site as in rule (2) we pick up a factor of

$$i(4+ma)^{-1}\tfrac{1}{2}(1+\gamma_\mu e_\mu) = iK_h(1+\gamma_\mu e_\mu). \qquad (10.45)$$

In a naive continuum limit the 'hopping constant' K_h goes to $1/8$. At finite lattice spacing, the critical value of K_h representing vanishing bare quark mass can be substantially renormalized through interactions. The strong coupling expansion is effectively in powers of β and K_h.

Equation (10.44) generalizes to all diagrams with the same topology as the diagram in figure 10.14, that is all diagrams with a single surface of plaquettes bounded by a quark line. This shows the striking connection between Wilson's theory and an oriented string model where the action

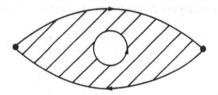

Fig. 10.16. A class of diagrams contributing to meson propagators
(Creutz, 1978a).

associated with a particular world sheet swept out by a string contains a term proportional to its area. In the strong coupling limit, the effective tension K in the string is the same quantity evaluated at the beginning of this chapter. In two-dimensional space-time this connection with the string model can be made precise (Bars, 1976). In four dimensions the picture is not the simple string model (Goddard *et al.*, 1973) due to complicated interstring interactions arising in higher orders (Weingarten, 1980).

The string analogy provides a useful topological classification of strong coupling diagrams. For example, a prototype diagram contributing to the pseudoscalar two-point function of eq. (10.41) is illustrated in figure 10.16. In this diagram the world sheet built up of plaquettes has a hole rimmed with a quark loop. The result for such a diagram is

$$-\tfrac{1}{3}3(4a^3+ma^4)^{-2}\,\mathrm{Tr}\,(\Gamma_E)\,\mathrm{Tr}\,(\Gamma_I)\,(4+ma)^{-p}\,(\beta/18)^4, \qquad (10.46)$$

where Γ_E is the product of the Dirac matrices around the external loop, and Γ_I is a similar product around the internal loop. Here the perimeter p is again the total quark line length and includes both the fermionic loops. The factor of $1/3$ in front of this expression represents the basis of the $1/n$ topological expansion (t'Hooft, 1974). Each additional quark loop inserted into a world sheet of a string will give another factor of $1/3$. In figure 10.17 we show a class of diagrams contributing to baryon structure. In strongly coupled lattice gauge theory, the proton consists of three quarks at the end of strings connected in a '*Y*' configuration.

As mentioned above, the strong coupling expansion is a simultaneous series in g_0^{-2} and K_h. In particular, the limit of infinite g_0^2 or vanishing β with K_h remaining finite is not exactly solvable. In this extreme, no plaquettes can be generated. Consequently, the two quarks of a meson must hop from

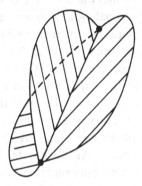

Fig. 10.17. A class of strong coupling diagrams contributing to baryon propagation (Creutz, 1978*a*).

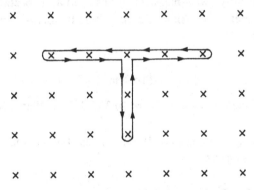

Fig. 10.18. A diagram giving a non-vanishing three-point function even in the limit of infinite g_0^2.

site to site together. Nevertheless, the hadrons are not free particles because they can still exchange these zero radius mesons. For example, we have non-vanishing three-point vertices of the type illustrated in figure 10.18. Although one should not expect a great deal of detailed phenomenological success, this limit has received considerable attention as an interesting simplification for the study of chiral symmetry breaking in a confining theory.

t'Hooft (1980) has suggested that there is an intimate connection between confinement and the phenomenon of chiral symmetry breaking. He argued that there are strong constraints which must be satisfied if a confining theory has massless bare fermion constituents and does not have massless Goldstone bosons associated with the chiral symmetry. The problem arises from the analytic structure of a three-point vertex constructed from two vector and one axial vector currents. The anomaly requires this object to be non-analytic at zero momentum transfer through the three channels. This requires real intermediate states of vanishing physical mass. In a confining theory these could be either Goldstone bosons or massless baryons. Further arguments (Coleman and Witten, 1980; Banks *et al.*, 1980) indicate the impossibility of the latter case in many theories, probably including the $SU(3)$ theory of the strong interactions.

This situation appears to carry over to the strongly coupled lattice theory. Such investigations require some treatment of the doubling problems alluded to in chapter 5. As the infinite g_0^2 theory with finite K_h is not exactly solvable, further approximations such as large dimension (mean field theory) or large gauge group are needed. The results of these calculations are strong indications that the theory adopts the broken symmetry alternative with massless 'pions' and an expectation value for the order parameter $\bar{\psi}\psi$ (Blairon *et al.*, 1981; Kluberg-Stern *et al.*, 1981; Svetitsky *et al.*, 1980).

Problems

1. Evaluate the diagram in figure 10.3 and show that it indeed gives a perimeter law.

2. Does strongly coupled $SO(3)$ lattice gauge theory confine in the sense of having a mass gap?

11

Weak coupling

Perturbation theory forms one of the mainstays in the development of modern theoretical particle physics. Indeed, the successes of perturbative quantum electrodynamics lie at the heart of our nearly universal adoption of renormalizable quantum field theory as the framework with which to describe elementary particle interactions. As our space-time lattice represents a regulator for ultraviolet divergences, in principle all perturbative results could be reproduced in this formalism. The basic expansion parameter g_0^2 represents the temperature in the analog statistical system. At low temperatures the important degrees of freedom are low energy excitations involving gentle long-wavelength variations of the fields. In magnetic systems the analogous excitations are referred to as spin waves and perturbation theory is a spin wave expansion.

Perturbative analysis did not motivate the original formulation of lattice gauge theory. Highly developed methods for calculation already exist for other cutoff schemes such as that of Pauli and Villars (1949) or dimensional continuation (Ashmore, 1972; Bollini and Giambiagi, 1972; t'Hooft and Veltman, 1972). Because of this, perturbation theory on a lattice has received little attention and remains rather awkward. In this short chapter we merely sketch spin wave techniques for lattice gauge theory. We will only evaluate the lowest order contribution to the average plaquette. It is somewhat ironic that this weak coupling regime has played such a minor role in lattice gauge theory and yet it is exactly this region to which we must go for a continuum limit, as will be discussed in the next chapters. The main virtue of the lattice remains in non-perturbative analysis.

We limit this discussion to the pure gauge theory with partition function

$$Z = \int (dU) \exp(-\beta \sum_{\square} (1 - (1/n) \operatorname{Re} \operatorname{Tr} U_{\square})). \tag{11.1}$$

As the inverse coupling β becomes large, this integral is increasingly dominated by $U_{\square}$ near the identity. Perturbation theory begins with a saddle point approximation taken at this maximum of the exponentiated action. We parametrize the plaquette operators

$$U_{\square} = \exp(i\lambda^\alpha \omega_{\square}^\alpha), \tag{11.2}$$

where the matrices λ^α generate the group and are normalized as in eqs (6.6–7). To leading order we have

$$1 - (1/n)\,\mathrm{Re\,Tr}\,U_\square = (1/(4n))\,\omega_\square^\alpha\,\omega_\square^\alpha + O(\omega_\square^4), \qquad (11.3)$$

and Z becomes

$$Z = \int (\mathrm{d}U)\exp\left(-(\beta/(4n))\,\omega_\square^\alpha\,\omega_\square^\alpha + O(\beta\omega^4)\right). \qquad (11.4)$$

For large β the exponential is highly suppressed unless

$$\omega = O(\beta^{-\frac12}) = O(g_0). \qquad (11.5)$$

Thus the ω^4 terms in eq. (11.4) are of order the coupling constant squared.

To proceed we would like to evaluate the leading behavior of the integral in eq. (11.4) in the Gaussian approximation. Here we encounter a technical difficulty in that the integrand is not damped in all directions when considered as a function of the link variables U_{ij}. Indeed, a gauge transformation can arbitrarily alter any given link and yet leave the action unchanged. Gauge fixing is an essential first step in the perturbative analysis. Our integrand receives a Gaussian damping only for those directions which do not represent gauge degrees of freedom.

The details of the gauge choice will be unimportant to the discussion here. One possibility is to set all timelike links to the identity, i.e. work in the 'temporal' gauge, and then on the spacelike surface $t = 0$ to do the additional gauge fixing necessary to eliminate the freedom of time-independent gauge transformations. If we now select any particular link, its value will be driven to the identity when β goes to infinity. There is a non-uniformity to this limit because links far from the hypersurface $t = 0$ are less constrained than those near it. For this technical reason we impose an infrared cutoff by working on a finite lattice.

After the gauge fixing, one quarter of the links are no longer variables. The remaining links are driven to the identity, about which we can expand

$$U_{ij} = 1 + i\lambda^\alpha\omega_{ij}^\alpha + O(\omega_{ij}^2), \qquad (11.6)$$

$$\omega_\square^\alpha = \sum_{ij\,\in\,\square} \omega_{ij}^\alpha + O(\omega^2). \qquad (11.7)$$

The integration measure in the vicinity of the identity takes the simple form

$$\mathrm{d}U_{ij} = (J + O(\omega_{ij}^2))\,d^{n_g}\omega_{ij}, \qquad (11.8)$$

where the weight J will ultimately be absorbed as an irrelevant constant. Here n_g is the number of group generators. Now the partition function assumes the form

$$Z = K\int \prod_{\{ij\}} d^{n_g}\omega_{ij}\exp\left[(-\tfrac12\beta\omega D^{-1}\omega) + O(\beta\omega^3)\right]. \qquad (11.9)$$

Here K is an overall constant factor and D^{-1} is a large matrix operating in the space of the variables ω_{ij}. In this form the partition function looks much like that discussed for a free field in chapter 4. The operator D is the propagator for the gauge gluons and enters into the Feynman diagrams of the theory. The $O(\beta\omega^3)$ terms are of order the coupling constant. They generate the vertices of the perturbative expansion.

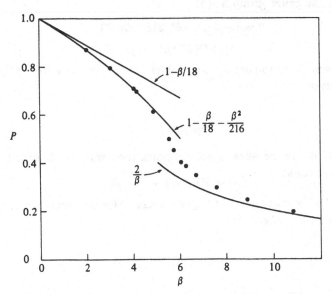

Fig. 11.1. The average plaquette for $SU(3)$ lattice gauge theory. The curves are the leading strong and weak coupling approximations and the points are from Monte Carlo analysis on 4^4 and 6^4 lattices.

For actual calculations these lattice propagators are quite cumbersome. However we can obtain some information on the average plaquette with very little effort. As our integral is now Gaussian, its value is a determinant

$$Z = K' |D/\beta|^{\frac{1}{2}} (1 + O(\beta^{-1})). \qquad (11.10)$$

The matrix D has the dimensionality of the parameter space after gauge fixing; consequently, it is a square matrix of $3n_g N^4$ rows. Here the factor of 3 is the number of non-fixed links per site. Removing a factor of β from each row of the matrix, we find

$$Z = K' |D|^{\frac{1}{2}} \beta^{-3n_g N^4/2} (1 + O(\beta^{-1})). \qquad (11.11)$$

For the average plaquette this implies

$$P = -(1/(6N^4))(\partial/\partial\beta) \log Z$$
$$= n_g/(4\beta) + O(\beta^{-2}). \qquad (11.12)$$

This result has a simple interpretation in statistical mechanics. We have $3n_g N^4$ physical variables distributed over $6N^4$ plaquettes. If we give each degree of freedom $\frac{1}{2}kT = 1/(2\beta)$ average energy, then we obtain exactly eq. (11.12). This simple counting of variables receives corrections at higher temperatures where the non-linear interactions come into play.

In figure 11.1 we summarize the leading strong and weak coupling results for the gauge group $SU(3)$

$$P = 1 - \beta/18 - \beta^2/216 + O(\beta^3)$$
$$= 2/\beta + O(\beta^{-2}). \tag{11.13}$$

The points in the graph are the true values for the plaquette from Monte Carlo analysis.

Problems

1. Show that in the weak coupling regime the parameter b_3 of the last chapter behaves as
$$b_3(\beta) = 3(1 - 4/\beta + O(\beta^2)).$$

2. What is the leading weak coupling behavior for the average plaquette in Z_2 lattice gauge theory?

12
Renormalization and the continuum limit

Regarding the lattice merely as an ultraviolet cutoff, ultimately we must consider the continuum limit. As when removing any regulator, observable quantities should approach their physical values. On the lattice, however, it is often convenient to measure dimensionful quantities, such as masses, in lattice units. For example, the mass of the first excitation in units of the spacing a gives the correlation length

$$\xi = (ma)^{-1}. \tag{12.1}$$

In the continuum limit m should remain finite while our yardstick of length a becomes singular. Thus we are interested in obtaining a divergent correlation length. In statistical mechanics language, this is the expected behavior at a second order phase transition. For a continuum limit of a field theory defined with a lattice cutoff, we should find the points in the coupling parameter space where the corresponding statistical model exhibits critical behavior. The needed critical phenomenon does not occur in the strong coupling region of lattice gauge theory. From eq. (10.23) we see that the correlation length goes to zero as β becomes small. To take a continuum limit we must search for second order phase transitions at intermediate and small coupling.

As soon as we begin discussing the removal of an ultraviolet cutoff, we must address the question of renormalization. Indeed, quantum field theory is notorious for the plethora of divergences which must be removed in calculations of physical observables. The bare charges and masses which appear in the Lagrangian are in general not well defined and need renormalization. The bare couplings acquire an implicit cutoff dependence chosen in such a manner that physical quantities have a finite limit when the cutoff is removed. For a well-defined renormalizable theory, this procedure should yield unique finite limits for all observables.

In general there are many possible renormalization schemes. In quantum electrodynamics one usually fixes the physical electron mass and the coefficient of the long-range Coulomb force. These parameters of the continuum theory determine the bare mass and charge when a cutoff is in place. In a confining theory, such as we want for the strong interactions, the choice is less obvious. One popular selection for non-perturbative studies of the pure gauge theory without fermions is the coefficient K of the Wilson loop area law, which equals the coefficient of the long-distance linear potential between external sources with quark quantum numbers. Another possible choice would be the mass of some physical bound state, such as the lightest glueball.

All of the quantities mentioned in the previous paragraph are defined in terms of long-range effects. This is clear for the long-distance potentials, but it also applies to a particle mass as this parameter determines how the particle propagates over extended distances. It is, however, often convenient to consider physical observables involving only finite length scales. For example, in traditional perturbative renormalization-group discussions one studies vertex functions in momentum space with all legs off-shell at some arbitrarily selected momentum scale μ. Alternatively, one might be interested in some interparticle force at a finite range r. By varying these parameters μ or r, one studies the interrelationships of physics on different length scales.

For now we will restrict our discussion to a theory, such as quarkless gauge theory, which has only one bare dimensionless coupling parameter, g_0. A general physical observable H is a function of the bare coupling as well as the cutoff scale of length a and the scale r on which H is defined

$$H = H(r, a, g_0(a)). \tag{12.2}$$

Here we have explicitly shown the cutoff dependence of the bare coupling $g_0(a)$. The precise form of this dependence depends on the details of the renormalization scheme. For simplicity, we assume that H is dimensionless; if it were not we could simply multiply by the appropriate power of r to make it so. For example, from an interparticle force $F(r)$ construct $H = r^2 F$.

As a becomes small and we approach the continuum limit, H should lose cutoff dependence. It should do this while retaining a non-trivial dependence on the scale r. This can only occur at special values of g_0 where critical behavior involving vastly different length scales occurs. To see this more explicitly, consider changing the cutoff by a factor of two. For small cutoff H should not change appreciably if g_0 is appropriately adjusted

$$H(r, \tfrac{1}{2}a, g_0(\tfrac{1}{2}a)) = H(r, a, g_0(a)) + O(a^2). \tag{12.3}$$

In general there are two classes of dimensional parameters which set the scale for the order-a^2 corrections in this equation. First, of course, is the scale r used to define H. In addition, however, we must consider the long-range physical parameters characterizing the continuum theory. In particular, regardless of how large r is, we must expect corrections of order a^2m^2 where m is some typical mass in the physical particle spectrum. The lattice theory should only be expected to approximate continuum physics when the lattice spacing is smaller than both the scale under consideration and the characteristic size of a strongly interacting particle. Of course, if we adopt the renormalization scheme of holding $H(r, a, g_0(a))$ fixed at the given scale, then by definition there are no corrections to eq. (12.3). However, we will now consider varying r in order to compare physics on different length scales and therefore we should remember that these corrections are in principle there.

Since H is dimensionless, we can scale a factor of two from both r and a in eq. (12.3) to give

$$H(2r, a, g_0(\tfrac{1}{2}a)) = H(r, a, g_0(a)) + O(a^2). \tag{12.4}$$

This equation shows the correlation between the bare coupling for two values of the cutoff and the measured observable at two different length scales. The process leading to this result is now iterated to give the pivotal relation

$$H(2r, a, g_0(a/2^{n+1})) = H(r, a, g_0(a/2^n)) + O(a^2). \tag{12.5}$$

This formula allows us to study the renormalization of g_0 as follows. Assume that for some fixed values of r and a we know the functional dependence of $H(r, a, g_0)$ and $H(2r, a, g_0)$ on the bare coupling. Suppose further that at scale r and in the continuum limit H has the value H_0:

$$\lim_{a \to 0} H(r, a, g_0(a)) = H_0. \tag{12.6}$$

Consider a graph of $H(r, a, g_0)$ as a function of g_0. Neglecting finite cutoff corrections, we find $g_0(a)$ as the value of g_0 where H passes through H_0. Now from $H(2r, a, g_0)$ we find the bare coupling at half this cutoff using eq. (12.4)

$$H(2r, a, g_0(\tfrac{1}{2}a)) = H_0. \tag{12.7}$$

Once we know $g_0(\tfrac{1}{2}a)$, we define H_1 by

$$H_1 = H(r, a, g_0(\tfrac{1}{2}a)). \tag{12.8}$$

Equation (12.5) now tells us how to find $g_0(\tfrac{1}{4}a)$:

$$H(2r, a, g_0(\tfrac{1}{4}a)) = H_1. \tag{12.9}$$

Iterating gives
$$H_n = H(r, a, g_0(a/2^n)), \tag{12.10}$$

$$H(2r, a, g_0(a/2^{n+1})) = H_n. \tag{12.11}$$

Graphically, this procedure generates a 'staircase' as illustrated in figure 12.1. This picture is drawn for an asymptotically free theory where $g_0(0) = 0$.

In figure 12.2 we sketch a situation where the functions $H(r, a, g_0)$ and

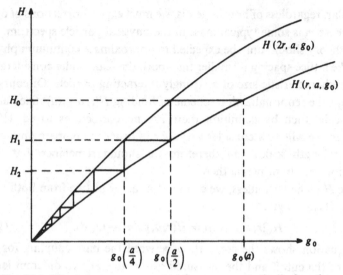

Fig. 12.1. The staircase construction for an asymptotically free theory (Creutz, 1981a).

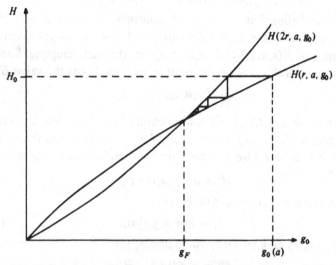

Fig. 12.2. An example of a non-trivial fixed point (Creutz, 1981a).

$H(2r, a, g_0)$ cross each other at a non-vanishing coupling. Here the staircase asymptotically approaches this crossing point. At this renormalization-group fixed point g_F, physics becomes scale invariant

$$H(r, a, g_F) = H(2r, a, g_F). \qquad (12.12)$$

Note that g_F can be approached either from stronger or weaker coupling.

As the bare charge at some very small cutoff passes through g_F, the corresponding initial value H_0 drastically changes as we go from a staircase on one side of g_F to the other. Long-distance physics depends non-analytically on the bare coupling and we have a phase transition in the corresponding statistical mechanical system. The critical exponents of the transition are related to the relative slopes of $H(r, a, g_0)$ and $H(2r, a, g_0)$ near the critical point. The absolute slopes of these functions depend on the initial value of a/r used in their definition.

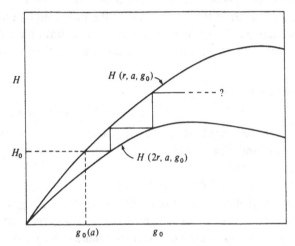

Fig. 12.3. A theory without a non-trivial continuum limit.

The above examples represent conventional ultraviolet attractive fixed points. One could also imagine a situation where at some point g_F eq. (12.12) again holds but

$$(|(d/dg)H(r, a, g)| - |(d/dg)H(2r, a, g)|)|_{g - g_F} > 0. \qquad (12.13)$$

In this case the staircase construction leads one away from g_F. A continuum limit at such an ultraviolet repulsive fixed point is at best possible only if g_0 is exactly g_F.

Another possible situation is that at some stage in the renormalization process eq. (12.11) has no solution. Such a case is illustrated in figure 12.3. At a certain point in the construction it is no longer possible to maintain H at its desired physical value regardless of what goes to the bare charge. Several authors (Kogut and Wilson, 1974; Baker and Kincaid, 1979; Bender *et al.*, 1981; Freedman, Smolensky and Weingarten, 1982) have suggested that this may be the case for four-dimensional ϕ^4 theory, which may therefore not have a non-trivial continuum limit.

The flows of the coupling illustrated in the above discussion represent a simplified version of the multidimensional flows discussed by Wilson (1971*a, b*). In particle physics we are usually interested in the continuum limit of a theory specified by a few renormalizable coupling constants. In statistical mechanics, however, the above rescaling procedure is often discussed in reverse. Starting with a simple model on a lattice of small spacing, one attempts to find an effective theory on a larger lattice spacing but with equivalent physics on long length scales. In general an increasing number of parameters is needed as such a process is iterated.

The above discussion of the dependence of the bare coupling on cutoff is often formulated in differential form. If our renormalization prescription is to set H at scale r to H_0 for all values of the cutoff a, then we have the equation

$$a(\mathrm{d}/\mathrm{d}a)H(r, a, g_0(a)) = 0$$
$$= a(\partial/\partial a)H(r, a, g_0) + \gamma(g_0)(\partial/\partial g_0)H(r, a, g_0). \quad (12.14)$$

This is a form of the renormalization group equation (Gell-Mann and Low, 1954; Petermann and Stueckelberg, 1953). The renormalization group function $\gamma(g_0)$ is defined

$$\gamma(g_0) = a(\mathrm{d}/\mathrm{d}a)g_0(a). \quad (12.15)$$

Knowledge of $\gamma(g_0)$ determines the cutoff dependence of g_0 up to an integration constant. Notice that once a renormalization prescription has been selected, then g_0 and a are no longer independent variables. We can freely trade off cutoff dependences for dependence on g_0 and vice versa. This interplay between dimensionful and dimensionless parameters forms the basis of the phenomenon of dimensional transmutation, the subject of the next chapter.

Zeros in the renormalization group function $\gamma(g_0)$ correspond to the scale-invariant crossing points discussed earlier in this chapter. As the lattice spacing becomes small, the bare coupling approaches a fixed point

$$\lim_{a \to 0} g_0(a) = g_F. \quad (12.16)$$

Equation (12.15) then implies

$$\gamma(g_F) = 0. \quad (12.17)$$

Note furthermore that for g_0 near g_F we have

$$\gamma(g_0) = a(\mathrm{d}/\mathrm{d}a)g_0(a) \begin{Bmatrix} > 0, g_0 > g_F \\ < 0, g_0 < g_F. \end{Bmatrix} \quad (12.18)$$

Thus for an ultraviolet attractive fixed point, such as being considered here, the first non-vanishing derivative of γ must be positive.

In general the precise form of γ will vary with the details of the renormalization scheme. In particular, γ depends on the choice of physical observable H and the scale r on which it is measured. Nevertheless, those zeros of γ representing ultraviolet-attractive fixed points must be universal if the continuum limit is to be unique. The scheme dependence of the renormalization group function appears already in the strong coupling limit. First consider the renormalization prescription of holding the string tension fixed. Application of eq. (10.9) to $SU(3)$ gives the strong coupling expression

$$K = a^{-2}\log(3g_0^2(a)) + O(g_0^{-2}). \tag{12.19}$$

If K is independent of a, a derivative gives

$$0 = a(\mathrm{d}/\mathrm{d}a)K = -2K + (2/(a^2 g_0))\gamma(g_0) + \ldots \tag{12.20}$$

Using eq. (12.19) to eliminate a^2 in favor of g_0, we have

$$\gamma(g_0) = g_0\log(3g_0^2) + \ldots \tag{12.21}$$

Note that this does not vanish in strong coupling; therefore, one must look elsewhere for a continuum theory.

Now suppose that, instead of using the string tension, we renormalize by holding the mass gap fixed. Equation (10.23) gives

$$m_g = a^{-1}4\log(3g_0^2) + O(g_0^{-2}). \tag{12.22}$$

Proceeding in analogy with eqs (12.19–21), we find

$$0 = a(\mathrm{d}/\mathrm{d}a)m_g = -m_g + (8/(ag_0))\gamma(g_0) + \ldots \tag{12.23}$$

$$\gamma(g_0) = \tfrac{1}{2}g_0\log(3g_0^2) + \ldots \tag{12.24}$$

Note the change in the normalization between eqs (12.21) and (12.24). Away from a zero of $\gamma(g_0)$ the lattice spacing is not small. This influences the relationships among observables and can appear as a scheme dependence of the renormalization-group function.

Problems

1. What does it physically mean to change the initial value H_0 in eq. (12.7)?

2. Suppose near a fixed point g_F that the renormalization-group function behaves as $\gamma(g_0) = (g_0 - g_F)\lambda + O((g_0 - g_F)^2)$. Show that the correlation length diverges at g_F as

$$\xi \propto (g_0 - g_F)^{-1/\lambda}.$$

13
Asymptotic freedom
and dimensional transmutation

In this chapter we return to the weak coupling limit of non-Abelian gauge theories. At the level of tree Feynman diagrams, relativistic field theory has no divergences and thus needs no renormalization. The bare coupling acquires cutoff dependence only after divergent one-loop diagrams are encountered. This implies that in the perturbative limit of our gauge theory of quarks and gluons

$$\gamma(g_0) = O(g_0^3). \tag{13.1}$$

At the outset we know that one zero of the renormalization group function occurs at vanishing coupling. For this root to be ultraviolet attractive and therefore useable for a continuum limit requires a positive sign for the first non-vanishing term in this perturbative expansion. Politzer (1973) and Gross and Wilczek (1973a, b) first calculated the relevant term for non-Abelian gauge theories. Defining the coefficients γ_0 and γ_1 from the asymptotic series

$$\gamma(g_0) = \gamma_0 g_0^3 + \gamma_1 g_0^5 + O(g_0^7), \tag{13.2}$$

we have the result for $SU(n)$ gauge theory with n_f fermionic species

$$\gamma_0 = (1/16\pi^2)(11n/3 - 2n_f/3). \tag{13.3}$$

Thus as long as

$$n_f < 11n/2, \tag{13.4}$$

the fixed point at the origin can potentially give a continuum limit. The two-loop contribution (Caswell, 1974; Jones, 1974) is

$$\gamma_1 = (1/16\pi^2)^2(34n^2/3 - 10nn_f/3 - n_f(n^2-1)/n). \tag{13.5}$$

Although in general $\gamma(g_0)$ is scheme dependent, these first two terms in its perturbative expansion are not. Consider two different schemes both defining a bare coupling as a function of cutoff: $g_0(a)$ and $g_0'(a)$. In the weak coupling limit each formulation should reduce to the classical Yang–Mills theory, and thus to lowest order they must agree

$$\left. \begin{aligned} g_0' &= g_0 + cg_0^3 + O(g_0^5), \\ g_0 &= g_0' - cg_0'^3 + O(g_0'^5). \end{aligned} \right\} \tag{13.6}$$

We now calculate the new renormalization group function

$$\gamma'(g_0') = a(\mathrm{d}/\mathrm{d}a)\,g_0' = (\partial g_0'/\partial g_0)\,\gamma(g_0)$$
$$= (1 + 3cg_0^2)\,(\gamma_0 g_0^3 + \gamma_1 g_0^5) + O(g_0^7)$$
$$= \gamma_0 g_0'^3 + \gamma_1 g_0'^5 + O(g_0'^7). \tag{13.7}$$

To order $g_0'^5$ all dependence on the parameter c cancels.

Thus far our discussion of the renormalization group has been in terms of the bare charge in the theory with a cutoff in place. This is a natural procedure in the lattice theory; however, the renormalization group is still useful in the continuum theory if we define a finite renormalized coupling constant. Like the generic physical function H of the last chapter, a renormalized coupling is first of all an observable which remains finite in the continuum limit

$$\lim_{a \to 0} g_R(r, a, g_0(a)) = g_R(r). \tag{13.8}$$

In general, the renormalized coupling g_R retains a dependence on the scale r of its definition. The masses and radii of the physical particles determine the typical dimensions for this dependence.

Secondly, to be properly called a renormalization of the classical coupling, g_R should be normalized such that it reduces to the bare coupling in lowest order perturbation theory when the cutoff is still in place.

$$g_R(r, a, g_0) = g_0 + O(g_0^3). \tag{13.9}$$

Beyond this, the definition of g_R is totally arbitrary. In particular, given any physical observable H satisfying the perturbative expansion

$$H(r, a, g_0) = h_0 + h_1 g_0^2 + O(g_0^4), \tag{13.10}$$

we can define a renormalized coupling

$$g_H^2(r) = (H - h_0)/h_1. \tag{13.11}$$

For perturbative purposes one often uses a renormalized three-gluon vertex with all legs at a given scale of momentum, representing the inverse of the scale r, and with a gauge fixing imposed.

In the continuum limit it should be possible to re-express physical observables such as H in terms of renormalized quantities. The renormalized perturbation expansion then takes the form

$$H(r, r', g_R(r')) = h_0 + h_1 g_R^2 + O(g_R^4). \tag{13.12}$$

Here r represents the scale on which H is defined and r' is the scale used to define the renormalized coupling. In general the coefficients in this series will differ from those in eq. (13.10); however to second order they agree. As r' is selected for convenience, changing its value should not alter real physical observables. This gives rise to the usual continuum renormalization-group equation

$$r'(\mathrm{d}/\mathrm{d}r')\,H(r,r',g_R(r')) = 0 = r'(\partial/\partial r')\,H + \gamma_R(g_R)\,H. \quad (13.13)$$

Here we have introduced the renormalized renormalization-group function

$$\gamma_R(g_R) = r(\mathrm{d}/\mathrm{d}r)\,g_R(r). \quad (13.14)$$

We can now draw a remarkable connection between this renormalization-group function and the one defined earlier for the bare coupling. When the cutoff is still in place, g_R is a function of the scale r, the cutoff a, and the bare coupling g_0

$$g_R = g_R(r, a, g_0).$$

However, since we are working with a dimensionless coupling, g_R can depend directly on r and a only through their ratio. This simple application of dimensional analysis implies

$$r(\partial/\partial r)\,g_R = -a(\partial/\partial a)\,g_R. \quad (13.15)$$

Now, as we renormalize the theory, g_R should become a function of r alone as a goes to zero, and we have

$$a(\partial g_R/\partial a) + (\partial g_R/\partial g_0)\,a(\partial g_0/\partial a) = 0. \quad (13.16)$$

Using the above equations and an analysis similar to that in eq. (13.7), we find

$$\gamma_R(g_R) = \gamma_0 g_R^3 + \gamma_1 g_R^5 + O(g_R^7). \quad (13.17)$$

The renormalized and bare γ functions have the same first two terms in their perturbative expansions. Indeed, it was through consideration of the renormalized coupling that γ_0 and γ_1 were first calculated.

Far from the weak coupling region, there is no simple relationship between the bare and renormalized γ functions. Perverse definitions (or not so perverse; see problem 1) of the renormalized coupling can lead to zeros in γ_R which have no counterpart in the bare quantities.

The perturbative expansion of γ_R has important experimental consequences. If we consider the continuum limit to be taken at $g_0 = 0$, and if g_R is ever small enough that the first terms dominate in eq. (13.17), then the renormalized coupling itself will be driven to zero as r becomes small. Not only does the bare coupling vanish, but any effective coupling becomes arbitrarily weak when the scale of measurement decreases. This is the physical implication of asymptotic freedom; phenomena involving only short-distance effects may be accurately described with the perturbative expansion. Indeed, asymptotically free gauge theories were first invoked for the strong interactions as an explanation of the apparently free parton behavior manifested in the structure functions of deeply inelastic scattering of leptons from hadrons.

Returning to the bare renormalization-group function, we wish to investigate how rapidly g_0 decreases with cutoff. Separating the variables in the form

$$\frac{dg_0}{\gamma_0 g_0^3 + \gamma_1 g_0^5 + O(g_0^7)} = d(\log a), \qquad (13.18)$$

we can integrate to obtain the result

$$g_0^{-2} = \gamma_0 \log(a^{-2}\Lambda_0^{-2}) + (\gamma_1/\gamma_0)\log(\log(a^{-2}\Lambda_0^{-2}) + O(g_0^2)). \quad (13.19)$$

Here the parameter Λ_0 represents a constant of integration. This equation indicates the well-known logarithmic decrease of the coupling with scale. The subscript on Λ_0 is to remind us that it has been defined from the bare charge and with the Wilson lattice cutoff. For the renormalized coupling, this equation should be rewritten for g_R with the cutoff a replaced by r and with a possibly different integration constant Λ_R.

The constant appearing upon integration of the renormalization-group equation represents a yardstick for measurement of the scales of the strong interactions. Its value is scheme dependent as can be seen by considering two different bare couplings related as in eq. (13.6). From the analog of eq. (13.19) for g_0' with its own Λ_0', we see

$$\log(\Lambda_0'^2/\Lambda_0^2) = c/\gamma_0, \qquad (13.20)$$

where c is the parameter appearing in eq. (13.6). Thus, perturbation theory relates the values of Λ_0 in two different schemes. Furthermore, this requires only a one-loop calculation even though two loops were needed to define Λ_0 through eq. (19).

Hasenfratz and Hasenfratz (1980) were the first to perform the necessary one-loop calculations to relate Λ_0 and Λ_R. Defining the renormalized coupling from the three-gluon vertex in the Feynman gauge and with all legs carrying momentum $\mu^2 = r^{-2}$, they found

$$\Lambda_R/\Lambda_0 = \begin{Bmatrix} 57.5, SU(2) \\ 83.5, SU(3), \end{Bmatrix} \qquad (13.21)$$

for the pure gauge theory. Note that not only is Λ scheme dependent, but that different definitions can vary by rather large factors. The original calculation of these numbers was rather tedious, involving intermediate definitions of the coupling and evaluation of one-loop diagrams with the lattice regulator. These numbers have been verified with calculationally more efficient techniques based on a study of the quantum fluctuations around a slowly varying classical background field (Dashen and Gross, 1981). These calculations have been extended to other lattice actions and to theories with fermions (Weisz, 1981; Kawai, Nakayama and Seo, 1981).

We have been discussing the bare coupling as a function of the lattice spacing. A useful alternative considers the coupling as a parameter which determines the cutoff. Inverting eq. (13.19), we have

$$a = \Lambda_0^{-1}(g_0^2 \gamma_0)^{-\gamma_1/(2\gamma_0^2)} \exp\left(-1/(2\gamma_0 g_0^2)\right)(1+O(g_0^2)). \qquad (13.22)$$

Note the essential singularity at vanishing bare coupling. The perturbative renormalization group is about to give us non-perturbative information. Multiplying by the corresponding mass, we can obtain the weak coupling dependence of a correlation length on the lattice

$$ma = \xi^{-1} = (m/\Lambda_0)(g_0^2 \gamma_0)^{-\gamma_1/(2\gamma_0^2)} \exp\left(-1/(2\gamma_0 g_0^2)\right)(1+O(g_0^2)). \qquad (13.23)$$

If m is the mass of a physical particle and remains finite in the continuum limit, then its value in units of Λ_0 is given by the coefficient of the weak coupling dependence indicated in eq. (13.23).

For the above discussion we could elect to work with the correlation length between operators which select any desired set of quantum numbers, such as spin, parity, etc. Thus the mass of any particle in units of Λ_0 is the coefficient of the weak coupling dependence of some correlation function, as in eq. (13.23). Furthermore, Λ_0 is universal, determined solely by the initial cutoff scheme. It will drop out of any dimensionless ratio of masses, which is then determined uniquely by the theory. This brings us to the remarkable conclusion that for pure gauge fields the strong interactions have no free parameters. The cutoff is absorbed into $g_0(a)$, which in turn is absorbed into the renormalization-group dependence of eq. (13.23). The only remaining dimensional parameter is Λ_0, which merely sets the scale for all other masses. In a theory considered in isolation, one may define Λ_0 to be unity. Coleman and Weinberg (1973) have given this process, wherein a dimensionless parameter g_0 and a dimensionful one a manage to 'eat' each other, the marvelous name 'dimensional transmutation'.

In the theory including quarks, their masses represent new parameters. Indeed these are the only parameters in the theory of the strong interactions. In the limit where the bare quark masses vanish, referred to as the chiral limit, we return to a zero parameter theory. In this approximation to the physical world, the pion mass is expected to vanish and all dimensionless observables should be uniquely determined by the theory. This applies not only to mass ratios, such as of the rho mass to the proton, but as well to quantities such as the pion–nucleon coupling constant, once regarded as a parameter for a perturbative expansion. As the chiral approximation has been rather successful in the predictions of current algebra, we hope that eventually we may develop the techniques to calculate these quantities. If they seriously disagree with experiment, the theory is wrong because there are no parameters to adjust. Given a qualitative agreement, a fine tuning of the small quark masses should give the pion its mass and complete the theory.

The exciting idea of a parameter-free theory is sadly lacking from treatments of the other interactions such as electromagnetism or the weak force. There the coupling $\alpha = 1/137$ is treated as a parameter. One might optimistically hope for inclusion of the appropriate non-perturbative ideas into a grand unified scheme ultimately rendering α and the quark and lepton masses calculable.

The renormalization group is indeed a rich subject. We have only touched on a few uses which we will find valuable in later chapters. Perhaps the most remarkable result of this chapter is that a perturbative analysis of the renormalization-group function can give important non-perturbative conclusions, such as eq. (13.23).

Problems

1. Define $g_R^2(r)$ to be proportional to r^2 times the force between two quarks separated by a distance r. Argue that the corresponding renormalization-group function in the full theory of strong interactions including quark loops must exhibit a zero at non-vanishing g_R.

2. Show that the γ_1 term in eq. (13.19) is needed to properly define Λ_0.

14
Mean field theory

We have seen that confinement arises naturally in the strong coupling limit of the lattice theory, whereas the continuum limit with asymptotic freedom drives us toward the weak coupling regime. Desiring the qualitative features of confinement to persist in the continuum limit, we would like to be able to pass smoothly from high to low temperature in our statistical analog. This leads to the hope that $SU(3)$ lattice gauge theory has no phase transitions separating the strong and weak coupling domains.

Do we expect phase transitions in lattice gauge theory? In the chapter on discrete groups, we will show that indeed deconfining transitions do exist in some toy models. In this chapter we will present some non-rigorous arguments based on mean field theory which suggests that any gauge group potentially displays phase transitions in enough space-time dimensions. The approximation in mean field theory requires each variable to interact directly with a large number of neighbors; consequently, it is effectively a large dimension simplification.

The application of mean field theory to gauge systems has had a somewhat murky history. In its simplest form it ignores Elitzur's theorem, discussed in chapter 9. A link variable is assumed to have an expectation value, which is then calculated with a self-consistency condition. Rigorously, however, this expectation must vanish because the link is a gauge-variant object. Recent formulations (Drouffe, 1981; Flyvbjerg, Lautrup and Zuber, 1982) present mean field theory as a saddle point approximation which gives rise to a consistent expansion in inverse dimension. Gauge rotations appear as zero point modes in the first-order corrections and restore Elitzur's theorem. As our goal here is to motivate possible phase transitions in lattice gauge theory, we will present this approximation in a simple and heuristic form (Balian, Drouffe and Itzykson, 1974). In general, the mean field approach underestimates thermal fluctuations. For the second-order transitions in magnetic systems, the results overestimate critical temperatures, possibly by an infinite factor.

To emphasize the different predictions for spin and gauge models, we first illustrate the technique for the Ising model. Placing a spin s_i from

94

the set $Z_2 = \{1, -1\}$ on each site of a d-dimensional hypercubic lattice, we consider the partition function

$$Z = \sum_{\{s\}} \exp\left(\beta \sum_{\{ij\}} s_i s_j\right), \tag{14.1}$$

where each spin is summed over and the sum in the exponent is over all nearest-neighbor pairs of sites $\{ij\}$.

We wish to find an approximate expression for the magnetization, the expectation value for any given spin

$$M = \langle s_i \rangle. \tag{14.2}$$

We begin by considering a particular site i and replacing the spins on all neighboring sites with their average value M. Then the Boltzmann probability for the spin on site i to have value s_i becomes

$$P(s_i) = \exp(2d\beta M s_i)/(2\cosh(2d\beta M)), \tag{14.3}$$

where the factor $2d$ counts the number of neighbors to site i, and the denominator normalizes the probability. Requiring that the average value of s_i is also M, we obtain the self-consistency condition

$$M = \tanh(2d\beta M). \tag{14.4}$$

For small β, this equation has the unique solution $M = 0$. Mean field theory correctly predicts that the magnetization vanishes at high temperatures. In contrast, whenever

$$\beta > \beta_{mf} = 1/(2d) \tag{14.5}$$

eq. (14.4) also has a non-trivial solution with $M > 0$ (as well as a symmetric one at $M < 0$). A graphical solution of eq. (14.4) is illustrated in figure 14.1. To see that the latter solution is the favored one, consider eq. (14.4) iteratively. If initially M is slightly positive, the expectation of s_i will be increased and driven towards the non-vanishing solution. Thus mean field theory predicts a phase transition at β_{mf}. For larger β the system is predicted to spontaneously magnetize. In table 14.1 we compare this prediction with the known critical temperatures β_c for the Ising model in 1, 2, 3, and 4 dimensions (Fisher and Gaunt, 1964).

Table 14.1.

d	β_{mf}	β_c
1	0.500	∞
2	0.250	0.441
3	0.167	0.222
4	0.125	0.150

Note that the critical temperatures are always overestimated, for $d = 1$ by an infinite factor. The approximation does, however, improve as d increases.

This analysis predicts a continuous transition for the Ising model in any number of dimensions. As β decreases to β_{mf}, the non-trivial solution to eq. (14.4) decreases smoothly to zero. This will contrast sharply with the gauge theory, where all transitions are predicted to be first order; thermodynamics changes discontinuously at the phase transition.

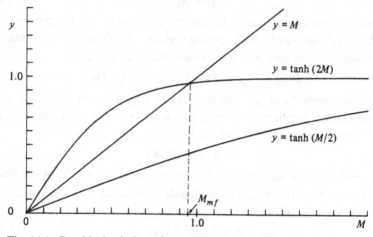

Fig. 14.1. Graphical solution of eq. (14.4) above and below the mean-field critical point.

In this example it was relatively easy to use physical arguments to determine which solution of eq. (14.4) was the relevant one. For the generalization to gauge theories it is useful to reformulate the technique in variational form (Peierls, 1938). For this purpose we use a convexity inequality on the exponential function. Given any function f over some set $X = \{x\}$ and a normalized measure $\rho(x)$ such that

$$\int_X \rho(x)\,\mathrm{d}x = 1, \qquad (14.6)$$

then, because the exponential function is convex, we have

$$\langle e^f \rangle \geqslant e^{\langle f \rangle}. \qquad (14.7)$$

Here the averages are with respect to the measure $\rho(x)$

$$\langle f \rangle = \int_X f(x)\rho(x)\,\mathrm{d}x. \qquad (14.8)$$

For the application of this inequality to the Ising system, we first add and subtract a source term to the action

$$Z = \sum_{\{s\}} \exp\left(\beta \sum_{\{ij\}} s_i s_j + H \sum_i s_i - H \sum_i s_i\right), \qquad (14.9)$$

where H will become a variational parameter. For our measure $\rho(x)$ we take

$$\int_X \rho(x)f(x)\,dx \to \sum_s \exp\left(H\sum_i s_i\right)f(s)/\left(\sum_s \exp\left(H\sum_i s_i\right)\right). \qquad (14.10)$$

Applying the convexity inequality with this measure gives

$$Z \geqslant \exp\left(N^d(d\beta\tanh^2(H) - H\tanh(H) + \log(2\cosh(H)))\right), \qquad (14.11)$$

where N^d represents the number of sites on our lattice. This translates into a bound on the free energy

$$\beta F = -N^{-d}\log(Z) \leqslant \beta F_{mf}(H)$$
$$= -d\beta\tanh^2(H) + H\tanh(H) - \log(2\cosh(H)). \qquad (14.12)$$

Minimizing the right hand side with respect to the parameter H optimizes the bound and gives rise to the relation

$$(d/dH)\beta F_{mf} = 0 = \mathrm{sech}^2 H(-2d\beta\tanh H + H). \qquad (14.13)$$

Note that this is equivalent to eq. (14.4) with the identification $H = 2d\beta M$. For low temperatures it is the non-zero root of this equation which minimizes the bound in eq. (14.12). In figure 14.2 we plot the mean-field free energy as a function of H for the cases $\beta = 1.1\beta_{mf}, \beta_{mf}$ and $\beta_{mf}/1.1$.

The terms in the mean-field free energy have a simple thermodynamical interpretation. The piece $-d\beta\tanh^2 H$ is a potential energy driving H to non-zero values. The remainder $H\tanh H - \log(2\cosh H)$ represents an approximation to an entropy factor trying to disorder the system. Which term wins depends on the value of β.

With this formalism in hand, we can proceed directly to the gauge theory. We consider the pure $SU(n)$ gauge theory normalized as in chapter 10 and study the partition function

$$Z = \int (dU)\exp\left((\beta/n)\sum_\square \mathrm{Tr}\,U_\square\right). \qquad (14.14)$$

Adding and subtracting $(H/n)\sum_{\{ij\}}\mathrm{Re\,Tr}\,U_{ij}$ from the action allows us to use the measure

$$\int_X \rho(x)f(x)\,dx \to \frac{\int(dU)\,e^{(H/n)\Sigma\,\mathrm{Re\,Tr}\,U}f(U)}{\int(dU)\,e^{(H/n)\Sigma\,\mathrm{Re\,Tr}\,U}} \qquad (14.15)$$

in the convexity inequality. This gives the bound

$$\beta F = -N^{-d} \log Z \leqslant \beta F_{mf}(H)$$
$$= -\tfrac{1}{2}d(d-1)\beta t(H)^4 + dHt(H) - d\log(c(H)). \qquad (14.16)$$

Here we have generalized the hyperbolic functions to arbitrary groups

$$c(H) = \int dH\, e^{(H/n)\,\mathrm{Re\,Tr}\,U}, \qquad (14.17)$$

$$t(H) = c(H)^{-1} \int dU \quad n^{-1}\,\mathrm{Re\,Tr}\,U\,e^{(H/n)\,\mathrm{Re\,Tr}\,U}. \qquad (14.18)$$

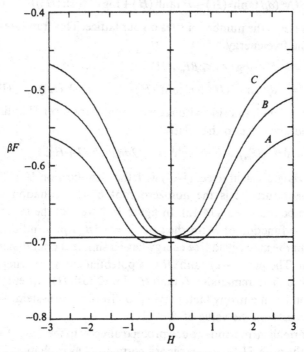

Fig. 14.2. The mean-field free energy for the Ising model. The curves *A*, *B*, and
C are for β 10% above, exactly on, and 10% below β_{mf}, respectively.

The factor $\tfrac{1}{2}d(d-1)$ in eq. (14.16) is the number of plaquettes per site on
a d-dimensional lattice. Note that for $SU(3)$ $t(H)$ is simply the parameter
$b_3/3$ of chapter 10, evaluated for $\beta = H$. Differentiating F_{mf} with respect
to H to find the extrema gives the consistency condition

$$(-2d(d-1)\beta t(H)^3 + dH)\, dt(H)/dH = 0 \qquad (14.19)$$

or

$$H = 2(d-1)\beta t(H)^3. \qquad (14.20)$$

The function $t(H)$ vanishes at $H = 0$. Thus $H = 0$ is always a solution of eq. (14.20). At high temperatures this is the only solution. At low temperatures further roots develop; however, in contrast to the Ising case, the root at the origin always represents a local minimum of F_{mf}. The potential term begins quarticly in H and thus the entropy piece will always dominate for small enough H. As β is increased another minimum develops at positive H. If β is large enough these new minima can be lower than

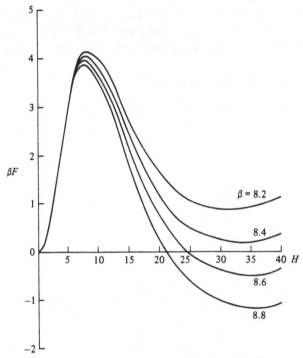

Fig. 14.3. The mean-field free energy as a function of H for several values of β with the group $SU(3)$ and $d = 4$.

the one at the origin. Here we see one dramatic difference from the Ising system; the transitions are predicted to be first order, with a discontinuous jump in the parameter H from zero to non-zero values. In figure 14.3 we plot the mean-field free energy as a function of H for several values of β for $SU(3)$ in four space-time dimensions.

Mean field theory predicts first-order phase transitions for all gauge groups. This prediction should only be trusted for large space-time dimensionality. If strong coupling arguments for confinement are to be relevant to the asymptotically-free continuum limit of lattice gauge theory, four dimensions must be inadequate for the mean field analysis to apply when the gauge group is $SU(3)$. In future chapters we will argue that this is the case. However, in four dimensions many gauge groups do give rise to phase transitions. Improved versions of mean field theory give rather accurate estimates for the transition temperatures (Flyvbjerg *et al.*, 1982).

Problems

1. Consider the partition function of eq. (14.1) with the variables s_i in the group $Z_3 = \{1, \exp(\pm 2\pi i)\}$ and a real part taken under the sum in the exponent. Show that mean field theory predicts a first-order transition for this model.

2. Prove the convexity inequality, eq. (14.7).

15

The Hamiltonian approach

We have been concentrating on the Euclidian path integral approach to lattice gauge theory. An alternative formulation, first advocated by Kogut and Susskind (1975), keeps a continuous time variable and only considers three-dimensional space as discrete. Working in the temporal gauge $A_0 = 0$, they define a Hamiltonian which is a function of the space components of the gauge field and a set of conjugate momenta. This formulation also permits a strong coupling expansion, which is now an application of quantum mechanical perturbation theory.

In this chapter we will derive the Kogut–Susskind Hamiltonian from the Wilson theory using the transfer matrix in direct analogy with the discussion in chapter 3. In this way we will see the equivalence of the two approaches. Which is preferable depends on taste and the particular question being asked. In the Wilson theory, space-time symmetry is more apparent, the particle spectrum is given by the singularity structure of Green's functions, and we have the simple analogy with statistical mechanics. In the Kogut–Susskind approach, we deal with a conventional quantum mechanical system with a well-defined Hamiltonian, the spectrum of the theory is directly the spectrum of this Hamiltonian, and phase transitions represent level crossings in the infinite volume limit.

As we wish to consider the continuous time limit of the Wilson theory, we introduce a different lattice spacing a_0 for the time direction. This gives the timelike plaquettes a different shape than the spacelike ones and the details of the argument in chapter 7 on the classical continuum limit must be slightly modified. The couplings on spacelike and timelike plaquettes are no longer equal in the action

$$S = -\beta_s \sum_s \text{Tr}\, U_\square - \beta_t \sum_t \text{Tr}\, U_\square. \tag{15.1}$$

Here the notation means that the first sum is over spacelike plaquettes only and the second over timelike ones. To obtain a proper classical limit we should take

$$\beta_s = 2na_0/(g_0^2 a), \tag{15.2}$$

$$\beta_t = 2na/(g_0^2 a_0), \tag{15.3}$$

101

where a continues to denote the spacelike lattice spacing. As a_0 goes to zero with fixed a, β_s goes to zero and β_t goes to infinity.

The above argument is essentially classical. For the quantum theory we have seen that the bare charge is cutoff-scheme dependent. In particular, the spacelike and timelike couplings may correspond to different Λ parameters in the sense discussed in chapter 13. Indeed, if we do not allow for such a change in relative spacelike and timelike scales, the speed of light may need to be renormalized (Shigemitsu and Kogut, 1981; Hasenfratz and Hasenfratz, 1981). Therefore we introduce two bare couplings and their geometric mean

$$\beta_s = 2na_0/(g_s^2 a), \qquad (15.4)$$

$$\beta_t = 2na/(g_t^2 a_0), \qquad (15.5)$$

$$g_H^2 = g_s g_t. \qquad (15.6)$$

The subscript on g_H stands for the Hamiltonian formulation. As with any bare couplings, these must all agree to lowest order

$$g_s^2 = g_t^2 + O(g_t^4) = g_H^2 + O(g_H^4). \qquad (15.7)$$

Introducing a cutoff dependence into the couplings and taking a continuum limit at the asymptotically free fixed point, we conclude

$$g_s^2(a)/g_t^2(a) \xrightarrow[a \to 0]{} 1. \qquad (15.8)$$

To proceed toward the Hamiltonian formulation, we now go to the temporal gauge. Fixing all timelike links to the identity, we see that a timelike plaquette represents a coupling between two spacelike links at subsequent times. Separating out time dependences, we relabel the sites with two indices, i and t, such that the first corresponds to the spatial coordinates and $a_0 t$ represents the time. In this notation the unfixed links carry a time index and two space indices $U_{ij,t}$. The pure gauge theory action is now

$$S = -(2a/(g_t^2 a_0)) \sum_{\{ij\}, t} \operatorname{Re} \operatorname{Tr}(U_{ij,t+1}^{-1} U_{ij,t}) - (2a_0/(g_s^2 a)) \sum_{\square, t} \operatorname{Re} \operatorname{Tr}(U_{\square,t}), \qquad (15.9)$$

where the second sum is over all spacelike plaquettes and all times.

In analogy with chapter 3 we wish to find a Hilbert space and an operator T such that

$$Z = \int (dU) e^{-S} = \operatorname{Tr} T^N, \qquad (15.10)$$

where N is the number of discrete times and we have imposed periodic boundary conditions. From the logarithm of T we will obtain the Hamiltonian. The first term in eq. (15.9) will generate the kinetic energy and the second, the potential.

The space in which T operates is a direct product of spaces of square-integrable functions over the gauge group. A state $|\psi\rangle$ in this space is specified by a wave function $\psi(U)$ which is a function of link variables U_{ij} which are group elements associated with each bond of a spacelike lattice. The inner product in this space is

$$\langle\psi'|\psi\rangle = \int(\mathrm{d}U)\,\psi^+(U)\,\psi(U). \tag{15.11}$$

For simplicity we use the same notation as for the path integral, but in eq. (15.11) only spacelike variables enter. We can expand the states of this space in the non-normalizable basis $\{|U\rangle\}$, where a state in this set is determined by a group element U_{ij} on each spacelike bond. These satisfy a condition that the reversed links are not independent

$$U_{ij} = U_{ji}^{-1}. \tag{15.12}$$

The overlap of states in this basis is

$$\langle U'|U\rangle = \prod_{\{ij\}} \delta(U'_{ij}, U_{ij}), \tag{15.13}$$

where the delta function over the group was introduced in chapter 9, eq. (9.14). The completeness statement is

$$1 = \int(\mathrm{d}U)|U\rangle\langle U|. \tag{15.14}$$

The general state takes the form

$$|\psi\rangle = \int(\mathrm{d}U)|U\rangle\,\psi(U). \tag{15.15}$$

Working in this Hilbert space, one may write down by inspection the matrix elements of an operator satisfying eq. (15.10)

$$\langle U'|T|U\rangle = \exp\left((2a/(g_t^2\,a_0))\sum_{\{ij\}}\mathrm{Re}\,\mathrm{Tr}\,(U'^{-1}_{ij}U_{ij})\right)$$
$$\times \exp\left((2a_0/(g_s^2\,a))\sum_{\square}\mathrm{Re}\,\mathrm{Tr}\,(U_\square)\right) \tag{15.16}$$

Just as we expressed T for quantum mechanics in terms of the operators $\hat{p}$ and $\hat{x}$, we would like to write this T in terms of some simple operators in the present Hilbert space. We begin by defining a set of matrix valued operators $\hat{U}_{ij}$ and unitary operators $R_{ij}(g)$, where g is an element of the gauge group

$$\hat{U}_{ij}|U\rangle = U_{ij}|U\rangle, \tag{15.17}$$

$$\left.\begin{aligned}R_{ij}(g)|U\rangle &= |U'\rangle,\\ U'_{ij} &= gU_{ij}\end{aligned}\right\} \tag{15.18}$$

and R_{ij} does not alter any other links. The operators $\hat{U}$ clearly are the analog of the coordinate $\hat{x}$ in ordinary quantum mechanics. The operators

$R_{ij}(g)$ satisfy the group representation property

$$R_{ij}(g) R_{ij}(g') = R_{ij}(gg').$$ (15.19)

They translate the variables U and thus are related to the canonical momentum, in a sense which will be made more precise shortly. In terms of these quantities, T takes the form

$$T = (\prod_{\{ij\}} (\int dg\, R_{ij}(g) \quad \exp((2a/(g_t^2 a_0))\, \text{Re Tr}\, g))$$
$$\times \exp((2a_0/(g_s^2 a)) \sum_{\square} \text{Re Tr}\, \hat{U}_{\square}),$$ (15.20)

where $\hat{U}_{\square}$ is the product of the $\hat{U}_{ij}$ around the corresponding plaquette.

We now wish to consider the limit as a_0 goes to zero. As a_0 becomes small, the integrals in eq. (15.20) become dominated by group elements near the identity. We parametrize the elements as in chapter 6

$$g = e^{i\omega^\alpha \lambda^\alpha} = e^{i\omega \cdot \lambda},$$ (15.21)

where $$\text{Tr}\,(\lambda^\alpha \lambda^\beta) = \tfrac{1}{2}\delta^{\alpha\beta},$$ (15.22)

$$[\lambda^\alpha, \lambda^\beta] = i f^{\alpha\beta\gamma} \lambda^\gamma.$$ (15.23)

The invariant group measure takes the form

$$dg = J(\omega) \prod_{\alpha} d\omega^\alpha.$$ (15.24)

The only properties of the Jacobean function J that we will need are that in a neighborhood of the identity it is regular and non-vanishing and that

$$J(\omega) = J(-\omega),$$ (15.25)

which follows because $dg = dg^{-1}$.

As with any representation of the group, the operator $R_{ij}(g)$ can be written in terms of a set of generators for that representation

$$R_{ij}(g) = \exp(i\omega^\alpha l_{ij}^\alpha) = \exp(i\omega \cdot l_{ij}).$$ (15.26)

In our Hilbert space the l_{ij}^α are Hermitian operators satisfying

$$[l_{ij}^\alpha, l_{ij}^\beta] = i f^{\alpha\beta\gamma} l_{ij}^\gamma,$$ (15.27)

$$[l_{ij}^\alpha, \hat{U}_{ij}] = -\lambda^\alpha \hat{U}_{ij},$$ (15.28)

$$[l_{ij}^\alpha, \hat{U}_{ji}] = \hat{U}_{ji} \lambda^\alpha,$$ (15.29)

$$[l_{ij}^2, l_{ij}^\alpha] = 0 = [l_{ij}^2, R_{ij}(g)].$$ (15.30)

The operators corresponding to different links all commute. In eq. (15.30) we have introduced the quadratic Casimir operator for the group

$$l_{ij}^2 = \sum_{\alpha} l_{ij}^\alpha l_{ij}^\alpha.$$ (15.31)

These operators may be all represented by differential operators in the group parameters. For example, with the group $U(1) = \{\exp(i\theta)\}$, we have a single generator $\lambda = 2^{-\frac{1}{2}}$ and

$$l_{ij} = 2^{-\frac{1}{2}}\mathrm{d}/\mathrm{d}\theta_{ij}. \tag{15.32}$$

To consider a link in the reversed direction, first note that eq. (15.12) carries over to the operators

$$\hat{U}_{ij} = \hat{U}_{ji}^{\dagger}. \tag{15.33}$$

The connection between l_{ij} and l_{ji} follows from

$$R_{ji}(g)|U_{ij}\rangle = |U_{ij}g^{-1}\rangle$$
$$= |(U_{ij}g^{-1}U_{ij}^{-1})\,U_{ij}\rangle$$
$$= R_{ij}(U_{ij}g^{-1}U_{ij}^{-1})|U_{ij}\rangle. \tag{15.34}$$

This implies for the generators

$$l_{ji}^{\alpha}|U\rangle = -G(U_{ij})^{\alpha\beta}l_{ij}^{\beta}|U\rangle, \tag{15.35}$$

where $G(g)^{\alpha\beta}$ denotes the adjoint representation of the group

$$g^{-1}\lambda^{\alpha}g = G(g)^{\alpha\beta}\lambda^{\beta}. \tag{15.36}$$

As this is a real orthogonal representation, we have

$$l_{ij}^{2} = l_{ji}^{2}. \tag{15.37}$$

Thus the quadratic Casimir does not depend on the direction chosen for the link.

With this bit of group theory in hand, we return to the transfer matrix and insert eqs (15.21), (15.24) and (15.26) into eq. (15.20)

$$T = (\prod_{\{ij\}}(\int(\prod_{\alpha}\mathrm{d}\omega^{\alpha})\,J(\omega)\exp(il_{ij}\cdot\omega)\exp((2a/(g_t^2\,a_0))\,\mathrm{Tr}\cos(\omega.\lambda)))$$
$$\times\exp((2a_0/(g_s^2\,a))\sum_{\square}\mathrm{Re}\,\mathrm{Tr}\,U_{\square}). \tag{15.38}$$

When a_0 goes to zero, the integral over ω is dominated by ω near the maximum of $\mathrm{Tr}\cos(\omega\cdot\lambda)$. For a unitary group this maximum always occurs near $\omega = 0$. We expand about this point

$$\mathrm{Tr}\cos(\omega\cdot\lambda) = n - \tfrac{1}{4}\omega^2 + O(\omega^4). \tag{15.39}$$

Inserting this into eq. (15.38), we do the Gaussian ω integrals to obtain the result

$$T = K\exp(-a_0\,H + O(a_0^2)), \tag{15.40}$$

where K is an irrelevant constant factor and

$$H = (g_t/g_s)((g_H^2/(2a))\sum_{\{ij\}}l_{ij}^2 + (2/(g_H^2\,a))\sum_{\square}\mathrm{Re}\,\mathrm{Tr}\,\hat{U}_{\square}). \tag{15.41}$$

This is the Kogut–Susskind Hamiltonian.

The two terms in eq. (15.41) have a direct interpretation in analogy to the usual continuum gauge theory Hamiltonian. The second term is a sum over spacelike plaquettes and represents the lattice form of the magnetic field squared. The first term involves the canonical momenta and represents the electric field squared. Indeed, the operator l_{ij} corresponds directly to the flux of electric field passing through link ij.

In eq. (15.41) we have removed a factor of g_t/g_s so that the remainder of the Hamiltonian only depends on the mean, g_H. Note that by virtue of eq. (15.8), this prefactor approaches unity in the continuous space limit. Thus for spectrum calculations in the continuum, we can ignore this factor. The coupling g_H has its own associated Λ parameter, defined in analogy with eq. (13.19). As indicated there, the relationship of this parameter with any other scheme can be determined perturbatively. Hasenfratz and Hasenfratz (1981) have calculated

$$\Lambda_H/\Lambda_0 \begin{cases} = 0.84, \, n = 2 \\ = 0.91, \, n = 3. \end{cases} \qquad (15.42)$$

The above Hamiltonian possesses a large amount of symmetry due to the remaining gauge freedom of the theory. As we have only specified the temporal gauge, we can still do time-independent gauge transformations. An operator that performs such a transformation at space site i is

$$J_i(g) = \prod_{\{ij\} \supset i} R_{ij}(g), \qquad (15.43)$$

where the product extends over all bonds emanating from site i. This is a symmetry operator which commutes with the Hamiltonian. All physical states should be singlets under this operation in the sense that

$$J_i(g)|\psi\rangle = |\psi\rangle. \qquad (15.44)$$

In terms of the generators l_{ij}, this amounts to

$$\sum_{\{ij\} \supset i} l^\alpha_{ij}|\psi\rangle = 0. \qquad (15.45)$$

This equation says that the net electric flux out of any site is zero. Thus we have a discrete version of Gauss's law. Alternatively we could study external sources by allowing some sites to be other than a gauge singlet. Note that the counting of degrees of freedom parallels continuum treatments. The temporal gauge has removed timelike links as variables. Gauss's law removes one variable per group generator on each site. Thus the final theory has two degrees of freedom for each gauge boson, as expected from the possible polarizations in the continuum theory.

A strong coupling series is easily formulated for this Hamiltonian. When g_H is large, the electric term dominates. The kinetic part of the Hamiltonian is diagonalized by placing all links into singlet states with $l^2_{ij} = 0$. The natural basis of states for the strong coupling expansion is in terms of definite representations of the gauge group on each link. The potential or magnetic term in the Hamiltonian then acts as a perturbation which excites links into intermediate states involving higher representations. The first correction involves the excitation of the links around a single plaquette into the fundamental representation. For further details we refer the reader to the review by Kogut (1979).

Here we have only considered pure gauge fields. The Hamiltonian is easily extended to include fermionic or other matter fields. With fermions one again has the doubling problem alluded to in chapter 5 except that one factor of two is saved because time is continuous.

16
Discrete groups and duality

On a discrete space-time lattice the notion of continuity is lost. Remarkably, this gives us more freedom in formulating a gauge theory. Whereas classical continuum gauge fields require a continuous gauge group for non-triviality, this is no longer so in the Wilson theory. Indeed, it is straightforward to consider the lattice link variables to be elements of some finite discrete group. The simplest such model considers elements of the group $Z_2 = \{1, -1\}$ and represents a gauge-invariant interaction of a set of Ising spins. Wegner (1971) first introduced this system as an example of a case with non-trivial phase structure but without a local order parameter.

An amusing point is that although the classical theory based on a discrete group has no continuum limit, this does not necessarily carry over to the quantum theory. If the system has a second-order phase transition at an appropriate zero of its renormalization group function, one should be able to define a continuum quantum field theory.

One reason to study discrete variables is that the resulting models are often inherently more amenable to analysis. For example, the two-dimensional Ising model is exactly solvable for the thermodynamic functions, and yet it has a non-trivial ferromagnetic phase transition. A hope with gauge theories is to gain some insight into the nature of their phase structures. Furthermore, these models provide a useful testing ground for new techniques.

In this chapter we concentrate on the cyclic groups Z_P, where the elements are the P'th roots of unity. These models are all Abelian and as $P \to \infty$ we approach the $U(1)$ model. Obtaining the phase structure of the latter is essential because this is the gauge group of electrodynamics. Any attempt to understand quark confinement must also explain why QED, the prototype gauge theory, does not confine.

We shall use the Z_P gauge models in four dimensions as a framework for the discussion of duality transformations. This technique, which has also been extensively developed for spin systems (Savit, 1980; Cardy, 1980) relates the strong and weak coupling domains and in some cases determines phase transition temperatures exactly. Under duality, the thermodynamic functions of a model map onto a related system but with different couplings. Singularities either occur in dual pairs or are restricted to special self-dual points. Historically, this was first applied to the Ising model and gave an exact determination of the critical point (Kramers and Wannier, 1941). The extension to gauge theories is direct (Balian, Drouffe and Itzykson, 1975a; Korthals-Altes, 1978; Yoneya, 1978).

As in the usual formulation of lattice gauge theory, our variables are elements of the gauge group, which we take to be Z_P:

$$U_{ij} \in Z_P = \{e^{2\pi i k/P} \mid k = 0, ..., P-1\}. \tag{16.1}$$

Absorbing a factor of -1 in the action, the path integral is

$$Z = \sum_{U \in Z_P} e^{S(U)}. \tag{16.2}$$

Again as in the usual theory, the action is a sum over the plaquettes of the lattice

$$S(U) = \sum_{\square} S_{\square}(U_{\square}). \tag{16.3}$$

In the normal case $S_{\square}(U_{\square})$ is the real part of the trace of $U_{\square}$. We deviate slightly at this point because the discussion of duality is simplified if we consider a more general action per plaquette. To interpret e^S as a Boltzmann weight, we require $S_{\square}$ to be a real function. So that the orientation of the plaquettes is irrelevant, we also require

$$S_{\square}(U_{\square}) = S_{\square}(U_{\square}^{-1}). \tag{16.4}$$

Beyond these constraints, the action is arbitrary. For a general gauge group one usually requires that $S_{\square}$ is a class function over the group. For an Abelian group such as considered here all functions are class functions.

With any gauge group the general plaquette action has a character expansion

$$S_{\square}(U) = \sum_n \beta_n \chi_n(U), \tag{16.5}$$

where χ_n represents the trace in the n'th irreducible representation of the group. For Z_P there are precisely P such representations, all one-dimensional and given by

$$R_n(U) = U^n, \quad n = 0, ..., P-1. \tag{16.6}$$

The representation property

$$R_n(U) R_n(U') = R_n(UU') \tag{16.7}$$

is a trivial consequence of the Abelian nature of the group. To combine representations we have the rule

$$R_m(U) R_n(U) = R_{m+n}(U), \tag{16.8}$$

where the index $m+n$ is understood modulo P. The orthogonality of the characters is

$$P^{-1} \sum_U R_m(U) R_n(U) = P^{-1} \sum_U U^{m+n} = \delta_{m, P-n}. \qquad (16.9)$$

The crucial point which results in a simple duality structure for the Z_p theories is that the combination rule for characters has precisely the same structure as the original group. The dual theory will again be a Z_P theory. Duality has been much less useful in other theories, such as with non-Abelian groups, where the representation structure is more complicated.

With the above representations, eq. (16.5) becomes

$$S_\square(U) = \sum_n \beta_n U^n. \qquad (16.10)$$

In terms of these variables the constraint of eq. (16.4) becomes

$$\beta_n = \beta_{P-n}. \qquad (16.11)$$

The parameter β_0 is simply an overall normalization, irrelevant to thermodynamics but convenient to keep. Character orthogonality inverts eq. (16.5) with the result

$$\beta_n = P^{-1} \sum_U U^{-n} S_\square(U). \qquad (16.12)$$

For the discussion of duality it is convenient to expand the Boltzmann weight

$$e^{S_\square(U)} = \sum_n b_n U^n, \qquad (16.13)$$

where

$$b_n = P^{-1} \sum_U U^{-n} e^{S_\square(U)} \qquad (16.14)$$

and eq. (16.11) becomes

$$b_n = b_{P-n}. \qquad (16.15)$$

Up to a factor of b_0, these are the parameters which proved so useful in the strong coupling expansion. The energy shift represented by β_0 becomes an overall scale factor in the b_n. Thermodynamics depends on the latter parameters only in a projective sense.

To proceed, we take the path integral and insert the character expansion of eq. (16.14) for each plaquette. This gives a sum over an integer $n_\square$ associated with every elementary square of the lattice. Pulling this sum to the outside, the partition function is

$$Z = \sum_{\{n_\square\}} (\prod_\square b_{n_\square}) \prod_{\{ij\}} (\sum_{U_{ij}} (\prod_{\square \supset ij} U_{ij}^{n_\square})). \qquad (16.16)$$

Here the innermost product is over the six plaquettes containing the link ij. The sum over the U_{ij} is immediate from the orthogonality relations and gives

$$Z = \sum_{\{n_\square\}} (\prod_\square b_{n_\square}) \prod_{\{ij\}} (P \delta_{\Sigma_{\square \supset ij} n_\square, 0}), \qquad (16.17)$$

where the Kronecker delta is understood modulo P in its indices. The factor of P multiplying the delta functions occurs because we have not normalized our sums over the group. We now wish to make an appropriate change of variables which will enable us to do some of the sums with the Kronecker functions.

The factor $$\delta_{\Sigma_{\square \supset ij} n_{\square}, 0} \qquad (16.18)$$

involves the six plaquettes containing the link ij. In figure 16.1 we illustrate four of these, the remaining two utilize the unvisualized fourth dimension. The key to simplify this construct is to go to the dual lattice. We associate a new site with the center of each of the hypercubes of the original lattice.

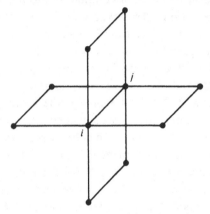

Fig. 16.1. Four of the six plaquettes containing the link ij. The remaining two utilize the fourth dimension.

For each site, link, plaquette, cube or hypercube on the original lattice there is a one-to-one correspondence with a hypercube, cube, plaquette, link or site, respectively, on the dual lattice. For example, dual to a link is the cube representing the common boundary of the two hypercubes which are dual to the ends of the link in question. Furthermore, the duality mapping can carry a sense of orientation if we invent a four-dimensional 'left hand' rule. For example, for the dual of a link in the positive t direction, we have a three-dimensional cube in xyz space. We can then adopt the convention that all plaquettes on this cube are oriented to the left when viewed from its center. For all other directions, we make even permutations on the indices x, y, z, and t. The dual of a plaquette is another plaquette, common to the four cubes which are dual to the links of the original plaquette. The orientation of the dual plaquette is specified by the above convention on any one of the original plaquette's links.

The utility of the dual mapping for the Z_P problem begins to appear with the observation that the six plaquettes needed in eq. (16.18) are dual to the set of six plaquettes which forms the three-dimensional cube dual to the link ij. Regarded on the dual lattice, the partition function is a sum over integers associated with each plaquette but subject to the constraint that the sum of these variables over any three-dimensional cube is zero modulo P, where the plaquettes are oriented with the appropriate handed rule. The next step is to solve this constraint.

If each of these dual plaquette variables were a modulo P sum of integers associated with the links of the dual lattice, then the constraint of eq. (16.18) would be automatically satisfied. This is because each link variable would occur twice, once in each orientation, in forming the cube variable and would thus cancel out. Remarkably, this solution of the constraint equation is unique up to gauge transformations. To see this, consider a completely fixed gauge in the sense discussed in chapter 9. Thus we set to zero all link variables on a maximal tree containing no closed loops. Any unfixed link must then form a unique closed loop with a set of fixed links. To solve the constraint, we set this link to the sum, modulo P, of the plaquette variables on any two-dimensional surface with this loop as its boundary. The constraint condition on cubes permits deformation of this surface and thus assures the uniqueness of the selection procedure. If we now undo the gauge fixing, we obtain P^{N^4} gauge equivalent configurations giving the dual plaquette variables as sums over the corresponding links.

This process eliminates the delta functions in eq. (16.17) and replaces the sum over plaquette variables with one over the dual link quantities

$$Z = P^{3N^4} \sum_{\{n_{ij}\}} \prod_{\square} b_{n_\square}. \tag{16.19}$$

Here $n_\square$ is the modulo P sum of the n_{ij} around the given plaquette. We now identify $n_\square$ with an element of Z_P in the natural way

$$U_\square = e^{2\pi i n_\square / P}, \tag{16.20}$$

and do a character expansion for b_n

$$b_{n_\square} = P^{-\frac12} e^{\tilde{S}(U_\square)} = P^{-\frac12} \sum_n \tilde{b}_n U_\square^n, \tag{16.21}$$

where this equation defines the dual action $\tilde{S}(U_\square)$. In terms of these new variables we reproduce the original partition function but with a new set of parameters $\tilde{b}_n$

$$Z(b) = Z(\tilde{b}). \tag{16.22}$$

The relation between b and $\tilde{b}$ is simply a linear transformation

$$\left.\begin{array}{l} \tilde{b}_n = A_{nm} b_m, \\ b_n = A_{nm}^* \tilde{b}_m, \end{array}\right\} \tag{16.23}$$

where A is the unitary matrix which generates discrete Fourier transforms

$$A_{nm} = P^{-\frac12} e^{2\pi i m n / P}. \tag{16.24}$$

This matrix has the properties

$$A^{-1} = A^* \quad \text{(unitarity)}, \tag{16.25}$$

$$A = A^T \quad \text{(symmetric)}, \tag{16.26}$$

$$(A^2)_{mn} = \delta_{m, P-n}, \tag{16.27}$$

$$A^4 = I. \tag{16.28}$$

Equation (16.22) is the key consequence of duality for the Z_P models. We note that the criterion of eq. (16.15) for orientation invariance automatically carries over to the dual variables because

$$A_{P-n, m} = A_{n, P-m}. \tag{16.29}$$

However, duality does not always result in a physically sensible model. If any of the $\tilde{b}_n$ are negative, then one cannot interpret them as new Boltzmann weights as in eq. (16.21). For those domains of the parameter space which are dual to another physical model, we have an interesting constraint on the singularities which can occur in the partition function. These must either occur in pairs, dual to each other, or must occur at self-dual points where $\tilde{b} = b$. There are examples of each of these possibilities.

To illustrate these ideas in a more specific case, we now turn to the Z_2 theory. Here the variables are from the set $\{1, -1\}$ and the action is

$$S_\square(U) = \beta_0 + \beta_1 U. \tag{16.30}$$

The parameters b_0 and b_1 follow from the expansion

$$\exp(\beta_1 U) = \cosh(\beta_1) + U \sinh(\beta_1). \tag{16.31}$$

This immediately gives

$$\left.\begin{array}{l} b_0 = \exp(\beta_0) \cosh(\beta_1), \\ b_1 = \exp(\beta_0) \sinh(\beta_1). \end{array}\right\} \tag{16.32}$$

Inverting these equations gives

$$\left.\begin{array}{l} \beta_0 = \tfrac{1}{2}\log(b_0^2 - b_1^2), \\ \beta_1 = \tfrac{1}{2}\log((b_0 - b_1)/(b_0 + b_1)). \end{array}\right\} \tag{16.33}$$

The Fourier matrix A is

$$A = 2^{-\frac{1}{2}}\begin{pmatrix} 1 & 1 \\ 1 & -1 \end{pmatrix}. \tag{16.34}$$

This gives the dual variables

$$\left.\begin{array}{l} \tilde{b}_0 = 2^{-\frac{1}{2}}(b_0 + b_1), \\ \tilde{b}_1 = 2^{-\frac{1}{2}}(b_0 - b_1). \end{array}\right\} \tag{16.35}$$

In terms of the variables β_0 and β_1 we have

$$\left.\begin{array}{l} \tilde{\beta}_0 = \beta_0 + \tfrac{1}{2}\log(\sinh(2\beta_1)), \\ \tilde{\beta}_1 = \tfrac{1}{2}\log(\tanh(\beta_1)). \end{array}\right\} \qquad (16.36)$$

The change in β_0 merely represents an overall normalization. The shift in β_1, however, represents a non-trivial change in the model. Small β_1 maps onto large $\tilde{\beta}_1$ and vice versa. Knowledge of the thermodynamic functions of the model in, say, the weak coupling regime determines, via eq. (16.22) and its derivatives, the corresponding functions in strong coupling. One point maps onto itself; this self duality occurs at

$$\beta_1 = \tfrac{1}{2}\log(1+2^{\frac{1}{2}}) = 0.4406867\ldots \qquad (16.37)$$

At exactly this coupling, numerical work has demonstrated that the model has a strong first-order phase transition, exhibited in chapter 9, figure 9.1 (Creutz, Jacobs and Rebbi, 1979a).

Returning now to general P, various contours in the multiparameter space reduce to standard models. The simple Wilson Z_P theory considers only β_0 and $\beta_{\pm 1}$. In this system for $P = 2$, 3, and 4, the model maps onto itself under duality. Monte Carlo analysis (Creutz, Jacobs, and Rebbi, 1979b) indicates strong first-order phase transitions at the self-dual points. At $P = 5$ or more, the model ceases to be exactly self-dual, the dual model requiring more than just β_0 and β_1. Numerical work on these models indicates two second-order phase transitions, one moving to larger β_1 as P increases, and the other remaining in the $U(1)$ limit. These features can be understood in terms of duality with a slightly modified one-parameter action which is self-dual.

The Villain (1975) variation of the Wilson theory considers the action

$$e^{S_\square(U)} = \sum_{l=-\infty}^{\infty} e^{-\frac{1}{2}\beta(\theta - 2\pi l)^2}, \qquad (16.38)$$

where the angle θ is defined

$$U = e^{i\theta}, \quad -\pi < \theta \leqslant \pi. \qquad (16.39)$$

For this action the parameters b_n are given by the double sum

$$b_n = P^{-1} \sum_{m=1}^{P} \sum_{l=-\infty}^{\infty} \exp(-\pi\beta(m/P - l)^2 - 2\pi i m n/P). \qquad (16.40)$$

To simplify this, we first complicate it by replacing the sum over l with an integral over a continuous angle and inserting a sum of delta functions from the formula

$$\sum_{l=-\infty}^{\infty} \delta(l - \theta/2\pi) = \sum_{k=-\infty}^{\infty} e^{ik\theta}. \qquad (16.41)$$

This gives

$$b_n = P^{-1} \sum_{m=1}^{P} \sum_{k=-\infty}^{\infty} \int_{-\infty}^{\infty} (d\theta/2\pi) \exp(-\tfrac{1}{2}\beta(\theta - 2\pi m/P)^2 - 2\pi i m n/P + ik\theta).$$

$$(16.42)$$

The theta integral is now Gaussian and yields

$$b_n = P^{-1}(2\pi\beta)^{-\frac{1}{2}} \sum_{k=-\infty}^{\infty} \sum_{m=1}^{P} \exp(-\tfrac{1}{2}k^2/\beta - 2\pi i m(n+k)/P). \quad (16.43)$$

The sum over m constrains $n+k$ to be zero modulo P. Thus we do this sum and replace the sum over k by a sum over multiples of P. This gives the final result

$$b_n = (2\pi\beta)^{-\frac{1}{2}} \sum_{k=-\infty}^{\infty} \exp(-\tfrac{1}{2}(P^2/\beta)(k-n/P)^2). \quad (16.44)$$

If we now return all the way back to eq. (16.21) and interpret this as the dual Boltzmann weight, we see that it has the same form as in eq. (16.38) but with β mapped onto

$$\tilde{\beta} = P^2/(2\pi\beta). \quad (16.45)$$

We have a self-dual model with the self-dual point at

$$\beta = (2\pi)^{-\frac{1}{2}}P. \quad (16.46)$$

As the parameter P goes to infinity the model goes over into $U(1)$. To avoid confinement in electrodynamics formulated with this lattice prescription, this model should exhibit a deconfining phase transition to a photon phase at weak coupling. Guth (1980) has rigorously proven the existence of such a transition. If the transition persists in the finite P models, then the latter must, by duality, have another transition at the dual point. This second transition, a consequence of the discreteness of the group, should move towards zero temperature with β growing as P^2 as P becomes large (Elitzur, Pearson, and Shigemitsu, 1979; Horn, Weinstein, and Yankielowicz, 1979; Ukawa, Windey and Guth, 1980). This is the empirically observed behavior for the Wilson theory, of which the Villain form is an approximation more amenable to analytic treatment.

Problems

1. Show that the two-dimensional Ising model is self-dual.
2. Show that the three-dimensional Z_2 gauge theory is dual to the three-dimensional Ising model.
3. Consider the P state gauge Potts model where all b_n except b_0 are equal to each other. Show that this model is self-dual. This system has a

single first-order phase transition at the self-dual point (Kogut, Pearson, Shigemitsu and Sinclair, 1980).

4. Show the self-duality of the Wilson Z_4 model with only the couplings β_0 and $\beta_1 = \beta_3$.

17

Migdal–Kadanoff recursion relations

In the chapter on mean field theory, we saw that with large space-time dimensionality, lattice gauge theories should exhibit first-order phase transitions. In contrast, in two dimensions the gauge systems reduce to simple one-dimensional spin chains (problem 1 of chapter 9) which exhibit no thermodynamical singularities. It is of crucial importance to know at what intermediate dimension the phase structure ceases to be non-trivial. Indeed, we want four space-time dimensions to be at or below this critical dimensionality for non-Abelian gauge groups so that we can use strong coupling techniques for the study of the confinement problem.

A similar question arises in spin models, where with large, such as three, dimensions ferromagnetic transitions occur whereas in one dimension long-range order is impossible without long-range forces. In the case of magnetic systems, there are rigorous theorems (Peierls, 1935; Mermin and Wagner, 1966; Coleman, 1973) which severely restrict the possible ordering in two dimensions for theories with a continuous symmetry group. The massless spin waves associated with a ferromagnetic state develop severe infrared singularities which disorder the system. Without a decoupling of the long-wavelength spin wave, which occurs in free field theory and with the $U(1)$ symmetry, these models cannot exhibit a massless phase. This gives us a compelling reason to believe that two is the critical dimensionality for spin systems with nearest-neighbor interactions. Indeed, ferromagnetism should not occur in these models when the symmetry is non-Abelian and the spin waves do not decouple. Unfortunately, we do not have such strong arguments for gauge theories.

From a renormalization group point of view, the phase structure of a theory appears when we compare the system on lattices with different spacing. As discussed in chapter 12, we have a second-order transition at a fixed point where physics does not change when we alter either the scale of measurement or the lattice spacing. The Migdal–Kadanoff recursion relations represent a simple approximate method for comparing theories with different lattice spacings (Migdal, 1975a, b; Kadanoff, 1976; 1977).

The virtue of this technique is that it provides a simple method for

117

obtaining an approximate renormalization group function. Its primary drawback lies in the difficulty of assessing the severity of the approximations involved. The procedure becomes exact in one or two dimensions for spin or gauge models, respectively. In contrast to mean field methods, the recursion should become less accurate as the dimension is increased. A particularly desirable feature of the method is that it correctly predicts two as the critical dimensionality for magnetic systems.

When applied to gauge theories in d dimensions, this approximate recursion gives precisely the same relation as for a spin model in $d/2$ dimensions. This immediately implies that four is the critical dimensionality for continuous gauge groups. In this sense, the absence of a phase transition in a four-dimensional non-Abelian gauge theory corresponds directly to the absence of ferromagnetism in two dimensions. Indeed, before the application of Monte Carlo methods to gauge systems, this was the strongest evidence for quark confinement in the standard gauge model of the strong interactions.

The d to $d/2$ correspondence between gauge and spin models predicts a similar structure for the $U(1)$ model of electrodynamics and the 'planar' or 'XY' model in two dimensions. In recent years the latter model has been the subject of considerable interest in solid state physics. It exhibits an infinite-order phase transition to a weak coupling phase with correlation functions which fall for large separations as a power of distance. This power is a continuously varying function of the coupling (Kosterlitz and Thouless, 1973). For the gauge theory of electrodynamics, this is consistent with the existence of a massless photon phase, which was mentioned in the last chapter. The renormalized electric charge is expected to be a continuously varying function of the bare coupling.

Although the Migdal–Kadanoff relations appear to correctly predict the critical dimensionality and the existence of at least some transitions, it can misidentify their nature. The Z_2 gauge model in four dimensions is predicted to be similar to the two-dimensional Ising model, whereas the former has a strong first-order transition and the latter, second order. For the $U(1)$ models discussed in the previous paragraph, the gauge model appears to be second order (Lautrup and Nauenberg, 1980a; DeGrand and Toussaint, 1980; Bhanot, 1981; Moriarty and Pietarinen, 1982) while the spin transition is of infinite order (Kosterlitz and Thouless, 1973).

We begin our detailed discussion with a demonstration of the technique on a trivial example, the one-dimensional Ising model. We consider a chain of N spins, each from the set $Z_2 = \{1, -1\}$. These interact through a nearest-neighbor coupling and give the partition function

$$Z = \sum_{\{s_i\}} \exp\left(\sum_{i=1}^{N} (\beta_0 + \beta_1 s_i s_{i+1}) \right). \tag{17.1}$$

As in the last chapter, we find it convenient to keep the normalization β_0 as a free parameter even if it is irrelevant to the thermodynamic singularities. For simplicity we treat the system as periodic, $s_{N+1} = s_1$. We now go to a transfer matrix formalism and write

$$Z = \text{Tr}(T^N). \tag{17.2}$$

Here T is the two-by-two matrix describing the interaction between neighboring spins

$$T = e^{\beta_0} \begin{pmatrix} e^{\beta_1} & e^{-\beta_1} \\ e^{-\beta_1} & e^{\beta_1} \end{pmatrix}. \tag{17.3}$$

In this example, the Migdal–Kadanoff relation merely represents an initial 'decimation' or sum over every other spin. In terms of the transfer matrix, we write (consider N even)

$$Z = \text{Tr}(T'^{N/2}), \tag{17.4}$$

where
$$T' = T^2 = e^{\beta_0'} \begin{pmatrix} e^{\beta_1'} & e^{-\beta_1'} \\ e^{-\beta_1'} & e^{\beta_1'} \end{pmatrix}. \tag{17.5}$$

The new couplings are given by

$$\beta_0' = 2\beta_0 + \tfrac{1}{2}\log\left(4\cosh\left(2\beta_1\right)\right), \tag{17.6}$$

$$\beta_1' = \tfrac{1}{2}\log\left(\cosh\left(2\beta_1\right)\right). \tag{17.7}$$

Thus the theory with parameters β_0 and β_1 has the same physics as the model on a lattice of twice the spacing but with different couplings β_0' and β_1'.

We can use this relation to take the lattice spacing to zero and obtain a continuum limit. This process requires repeatedly adjusting the couplings so as to maintain a constant correlation length in physical rather than lattice units. The parameter β_0 merely represents a zero point energy which cannot cause thermodynamic singularities. Thus we should concentrate on the physically relevant variable β_1. Equation (17.7) relates β_1 with cutoff a to its value with a lattice spacing of $2a$.

$$\beta_1(2a) = \tfrac{1}{2}\log\left(\cosh\left(2\beta_1(a)\right)\right). \tag{17.8}$$

In a continuum limit β_1 must go to a fixed point of this recursion relation. The two fixed points in this one-dimensional model are at $\beta_1 = 0$ and $\beta_1 = \infty$. The former of these is ultraviolet repulsive and the latter ultraviolet attractive in the sense discussed in chapter 12. This theory is asymptotically free in as much as we must go to infinite β_1 or zero 'temperature' for the continuum limit. This is the same behavior conjectured for the four-dimensional gauge theory of the strong interactions.

For further analysis, it is convenient to diagonalize T and find its eigenvalues. This is done directly via the character expansion of the exponentiated action in terms of the variables b_n of the last chapter.

$$T_{s,\,s'} = b_0 + b_1 ss', \tag{17.9}$$

where
$$\left.\begin{array}{l} b_0 = e^{\beta_0} \cosh(\beta_1), \\ b_1 = e^{\beta_0} \sinh(\beta_1). \end{array}\right\} \tag{17.10}$$

Up to a factor of two, these variables are the eigenvalues of T. The orthogonality of the characters now implies

$$(T^2)_{ss'} = 2(b_0^2 + b_1^2 ss'). \tag{17.11}$$

The factor of two arises because we have not normalized the sums over spins. Normally we set $\int dg = 1$, but here we have taken $\sum\limits_{s} 1 = 2$. The recursion relation for the variables b_i is thus a simple power

$$b_i' = 2b_i^2. \tag{17.12}$$

This generalizes to all groups. A decimation is a generalized convolution and becomes simple in the transform space of the characters. If the spins are considered as elements g of some group, we consider the partition function

$$Z = \int (\prod_i dg_i) \exp(\sum_i S_L(g_i g_{i+1}^{-1})), \tag{17.13}$$

where S_L is the contribution to the action from a single link. We now expand the nearest-neighbor interaction in characters of the irreducible representations of the group

$$\exp(S_L(g)) = \sum_R b_R \chi_R(g) = \exp(\sum_R \beta_R \chi_R(g)). \tag{17.14}$$

A decimation in this model will utilize the orthogonality

$$\int dg \chi_R^*(g) \chi_{R'}(gg') = d_R^{-1} \delta_{RR'} \chi_R(g'), \tag{17.15}$$

where d_R is the dimension of the matrices in the representation (problem 3 of chapter 8). This gives
$$b_R' = d_R^{-1} b_R^2. \tag{17.16}$$

We assume for a physical theory that b_R is real and that for every representation its conjugate occurs in eq. (17.14) with an equal coefficient.

Up to this point the recursion relations have been exact. In going to higher dimensions, approximations become necessary. Consider the two-dimensional Ising model with partition function

$$Z = \sum_{\{s_i\}} \exp\left(\sum_{\{ij\}} (\beta_0 + \beta_1 s_i s_j)\right), \qquad (17.17)$$

where the variables are again from Z_2 and the sum in the exponential is over all nearest neighbor pairs of sites on an N-by-N two-dimensional square lattice. We would like to begin with a decimation or sum over every other site in the x direction, for example those variables on sites with odd x coordinate. In figure 17.1 we show a portion of the lattice and label by

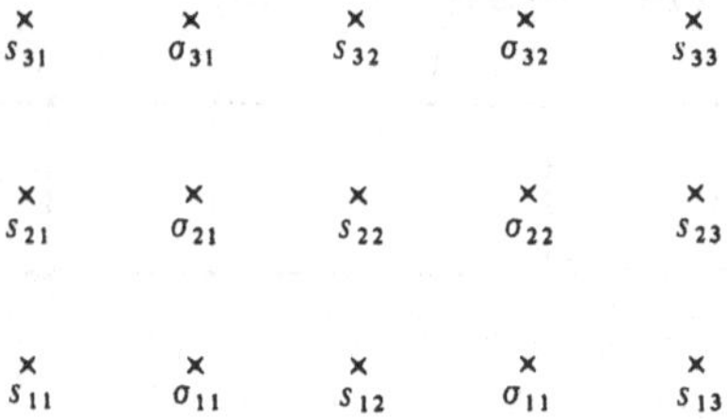

$$
\begin{array}{ccccc}
\times & \times & \times & \times & \times \\
s_{31} & \sigma_{31} & s_{32} & \sigma_{32} & s_{33} \\[1.5em]
\times & \times & \times & \times & \times \\
s_{21} & \sigma_{21} & s_{22} & \sigma_{22} & s_{23} \\[1.5em]
\times & \times & \times & \times & \times \\
s_{11} & \sigma_{11} & s_{12} & \sigma_{11} & s_{13}
\end{array}
$$

Fig. 17.1. A portion of a two-dimensional lattice. We wish to sum over the spins labeled by σ.

Fig. 17.2. A decimation generates non-nearest-neighbor couplings such as the diagonal bonds illustrated here.

σ those sites we wish to sum over. This should leave a model with action depending only on the remaining sites, labeled by s in the figure. Although in principle this can be done exactly, a complication arises because the new theory will in general involve non-nearest-neighbor interactions. This is schematically shown in figure 17.2. Indeed, the decimation will introduce couplings between spins on adjacent unsummed x rows but arbitrarily separated in the y direction. To keep the recursion relations manageable, some truncation is necessary. The simple Migdal–Kadanoff procedure eliminates these long distance interactions via the trick of bond moving.

The non-local couplings arise because the σ variables are coupled in the y direction. If these troublesome couplings were not present, then the sum over the σ would reproduce the one-dimensional recursion on the x direction bonds. The Migdal–Kadanoff approximation consists of neglecting the y couplings of the σ variables and compensating for them by increasing the strength of the y couplings between the unsummed s

variables. Effectively the bonds between the σ are moved one site over to give a double strength y bond between sites with even x coordinate. This bond moving is illustrated in figure 17.3. On the decimated lattice the new y coupling becomes

$$\beta_i^y \to 2\beta_i^y, \tag{17.18}$$

where the superscript indicates the direction associated with the coupling. The x bonds now receive the one-dimensional recursion from eqs (17.6) and (17.7)

$$\beta_0^x \to 2\beta_0 + \tfrac{1}{2}\log\left(4\cosh\left(2\beta_1^x\right)\right), \tag{17.19}$$

$$\beta_1^x \to \tfrac{1}{2}\log\left(\cosh\left(2\beta_1^x\right)\right). \tag{17.20}$$

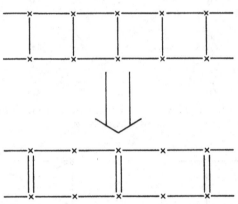

Fig. 17.3. Moving bonds to reduce the decimation to the one-dimensional case.

The next step is to repeat the decimation in the y direction and modify the couplings with x and y interchanged in the above equations. The net result is

$$\beta_0^y \to 4\beta_0^y + \tfrac{1}{2}\log\left(4\cosh\left(4\beta_1^y\right)\right), \tag{17.21}$$

$$\beta_1^y \to \tfrac{1}{2}\log\left(\cosh\left(4\beta_1^y\right)\right), \tag{17.22}$$

$$\beta_0^x \to 4\beta_0^x + \log\left(4\cosh\left(2\beta_1^x\right)\right), \tag{17.23}$$

$$\beta_1^x \to \log\left(\cosh\left(2\beta_1^x\right)\right). \tag{17.24}$$

Unfortunately the sequential decimation has lost the symmetry of the theory under interchange of the x and y axes. We will repair this momentarily but first note that this recursion does have a fixed point at the asymmetric couplings

$$\beta_1^x = 2\beta_1^y = 0.609\,377\,863\ldots \tag{17.25}$$

To recover the xy symmetry of the model we consider the respective decimations in an infinitesimal manner. Thus, we first perform the change of the lattice spacing in the x direction not by a factor of two, but by a factor of $(1+\Delta)$. The corresponding change in the x couplings is most apparent with the variables b_R, which become raised to the $(1+\Delta)$ power. For the general model, eq. (17.16) is replaced by

$$b'_R = d_R^{-\Delta} b_R^{1+\Delta} = b_R + b_R \log(b_R/d_R)\,\Delta + O(\Delta^2).\qquad(17.26)$$

For the y bond moving we increase the strength of the β_R^y by a factor of $(1+\Delta)$. Keeping only the leading terms in Δ, we repeat the decimation in the y direction. Both the x and y couplings change equally and become

$$\beta'_R = \beta_R + (\beta_R + \sum_{R'}(\partial\beta_R/\partial b_{R'})\,b_{R'}\log(b_{R'}/d_{R'}))\,\Delta.\qquad(17.27)$$

As the change in the lattice spacing is $a\Delta$, we obtain the Migdal–Kadanoff approximation to the renormalization group function

$$a\,\mathrm{d}\beta_R/\mathrm{d}a = \beta_R + \sum_{R'}(\partial\beta_R/\partial b_{R'})\,b_{R'}\log(b_{R'}/d_{R'}).\qquad(17.28)$$

The generalization to d dimensions is immediate; a decimation in any direction requires bond moving in all the $d-1$ orthogonal coordinates. The final formula is

$$a\,\mathrm{d}\beta_R/\mathrm{d}a = (d-1)\beta_R + \sum_{R'}(\partial\beta_R/\partial b_{R'})\,b_{R'}\log(b_{R'}/d_{R'}).\qquad(17.29)$$

Applied to the variable β_1 in the Ising model, this reduces to

$$a\,\mathrm{d}\beta_1/\mathrm{d}a = (d-1)\beta_1 + \sinh(\beta_1)\cosh(\beta_1)\log(\tanh(\beta_1)).\qquad(17.30)$$

At $d=2$, this has a fixed point at

$$\beta_1 = \tfrac{1}{2}\log(1+2^{\frac{1}{2}}),\qquad(17.31)$$

which remarkably is the exact result, as predicted by duality. As d goes to unity, the fixed point goes to infinity. Using the asymptotic forms for the hyperbolic functions, we obtain for the fixed point

$$\beta_1 = \tfrac{1}{2}(d-1)^{-1} + O(e^{-2\beta_1}).\qquad(17.32)$$

Thus we say that unity represents the critical dimensionality for the Ising model.

For a model based on a continuous group, the recursion relations predict two as the critical dimensionality. Physically this follows because at weak coupling the exponentiated action strongly peaks near the identity element of the group and approximates a Gaussian in the group parameter space. The decimation in the x direction convolutes these Gaussians, increasing their width by a factor of $(1+\Delta)$. In contrast, the bond moving decreases the widths of the Gaussians on orthogonal bonds by a factor $(1+\Delta)^{-1}$. For precisely two dimensions these operations are done equally on all bonds and the leading effects cancel.

To see this in more detail, it is convenient to introduce a new variable which simplifies the form of the recursion relation. We define

$$f_R = \log(b_R/(d_R b_0))$$
$$= \log(d_R^{-1}(\partial/\partial\beta_R)\log b_0), \qquad (17.33)$$

where b_0 is the b parameter for the singlet representation

$$b_0 = \int dg \exp(S_L(g)). \qquad (17.34)$$

The variable f_R has several nice properties. First, it is readily calculable from b_0 alone. Second, being a function of the ratio of two b parameters, it is independent of the overall normalization represented in β_0. Finally, the recurrence relation assumes a 'linearized' form

$$a\,df_R/da = (1 + (d-1)\sum_{R'} \beta_{R'}\,\partial/\partial\beta_{R'})f_R. \qquad (17.35)$$

We now consider the weak coupling limit of a truncated action with only a single coupling β representing the fundamental group representation. The parameter b_0 is then

$$b_0 = \int dg\, e^{\beta \operatorname{Re Tr}(g)}. \qquad (17.36)$$

From this we define the corresponding f variable

$$f = \log(d_F^{-1}(\partial/\partial\beta)\log(b_0)), \qquad (17.37)$$

where d_F is the dimension of the group matrix g. The recursion relation for f is

$$a(d/da)f = (1 + (d-1)\beta\partial/\partial\beta)f. \qquad (17.38)$$

If we now let β become large, the integral in eq. (17.36) receives its dominant contribution from g near the identity, where we write

$$g = e^{i\omega \cdot \lambda}, \qquad (17.39)$$

$$\operatorname{Re Tr}(g) = d_F - \tfrac{1}{4}\omega^2 + O(\omega^4). \qquad (17.40)$$

Here λ^α are the group generators, of which there are n_g. Straightforward Gaussian integration gives

$$f = n_g/(2\beta d_F) + O(\beta^{-2}). \qquad (17.41)$$

Inserting this into eq. (17.38) gives

$$a(d/da)\beta = (d-2)\beta + O(1), \qquad (17.42)$$

which shows the critical nature of two dimensions. The $O(1)$ term depends on the details of the quartic term in the action as well as the group measure. This sensitivity presumably is a signal of overextension of the approximations in the recursion relations; nevertheless, Kadanoff (1977) has given heuristic arguments which suggest a negative sign for this correction in two-dimensional non-Abelian theories. This supports the perturbative prediction of asymptotic freedom in these models (Polyakov, 1975).

With this machinery in hand, the generalization to gauge theories is direct. Indeed, this extension is almost trivial with the choice of an appropriate gauge. In d dimensions, to do a decimation along, say, the x axis, it is natural to work in the axial gauge where the links along that axis are all set to the identity. We then have a set of coupled one-dimensional chains of spins, as discussed in the chapter on gauge fixing. Suppose we now take those plaquettes which are transverse to the decimation direction and move those with odd x coordinate back one site. Then we can integrate the variables with odd x coordinate. In this procedure, the decimation on the plaquettes parallel to the x axis is precisely that of the one-dimensional spin system. After the decimation in one direction, we undo the gauge fixing along that axis and repeat the entire process along another. Continuing to make an infinitesimal decimation in every direction gives the Migdal–Kadanoff approximation for the renormalization group function in a gauge theory

$$a(\mathrm{d}/\mathrm{d}a)\beta_R = (d-2)\beta_R + 2\sum_{R'}(\partial\beta_R/\partial b_{R'})\, b_{R'}\log b_{R'}. \qquad (17.43)$$

The parameters β_R and b_R are defined in analogy with eq. (17.14) with S_L replaced by $S_\square$ and $g_i g_{i+1}^*$ replaced by $U_\square$. The factor $d-2$ in the first term arises because for any plaquette we perform bond moving for the $d-2$ dimensions orthogonal to that plaquette, while the other two dimensions in the plane of the plaquette give the factor of two in the second term.

This is the result advertised at the beginning of this chapter. Up to an overall factor of two, the recursion relation is identical to that for the spin system in $d/2$ dimensions. The correspondence appears because the spin interaction is along one-dimensional bonds while the gauge interaction utilizes two-dimensional plaquettes. The important prediction is that the critical dimensionalities in the gauge theory are twice those of the spin models. Thus our four-dimensional world represents a critical case for continuous gauge groups.

Problems

1. As d goes to infinity, what happens to the fixed point of eq. (17.30)? Which should be more reliable, this prediction or that of mean field theory?

2. Consider an action with a quartic term

$$S_L(g) = (d_F + \tfrac{1}{4}\omega^2 + C\omega^4 + O(\omega^6))$$

in the analysis leading to eq. (17.42). Find the $O(1)$ terms in the latter equation in terms of the parameters C and the integration measure $\mathrm{d}g = \mathrm{d}^n{}_s\omega\,(J_0 + J_1\omega^2 + O(\omega^4))$.

3. Consider the general Z_P model discussed in the last chapter. Show that the processes of bond moving and decimation interchange under duality. Thus the infinitesimal recursion relation respects the duality symmetry and gives the exact result in eq. (17.31).

4. What does the Migdal–Kadanoff relation predict for the behavior of the correlation length near the critical point of the two-dimensional Ising model?

18

Monte Carlo simulation I: the method

The lattice formulation reduces the Feynman path formula for the gauge theory into a multiple ordinary integral. This suggests that, at least for finite size systems, one might attempt to numerically evaluate the partition function. A moments thought, however, reveals that the high multidimensionality of the integrals makes conventional mesh techniques impractical. For example, consider a 10^4 site lattice, a size fairly typical for numerical work. Such a system has 40 000 link variables. If we now take the simplest possible gauge theory, that with gauge group Z_2, the partition function becomes an ordinary sum. But this sum has an enormous number of terms, that number being

$$2^{40\,000} = 1.58 \times 10^{12\,041}. \qquad (18.1)$$

Even if we could add one term in the time it takes light to pass by a proton and continue for the age of the universe, we would not put a perceptible dent in the sum.

The appearance of such large numbers immediately suggests a statistical treatment. Indeed, there are also an enormous number of ways to place molecules of H_2O into a glass and yet one only needs a few to determine the thermodynamic properties of water. The goal of the Monte Carlo approach is to provide a small number of configurations which are typical of thermal equilibrium in the statistical analog. Whereas the super-astronomical number of terms indicated in eq. (18.1) can never be summed exactly, it is straightforward to store the few tens of thousands of numbers characterizing typical configurations which strongly dominate the sum.

A Monte Carlo program begins with some initial configuration of the fields stored as an array in a computer memory. It then sequentially makes pseudo-random changes on these variables in such a manner that the ultimate probability density for encountering any configuration C is proportional to the famous Boltzmann factor

$$p_{eq}(C) \propto e^{-\beta S(C)}, \qquad (18.2)$$

where $S(C)$ is the action associated with the given configuration. In this chapter, to emphasize the connection with statistical mechanics, we

127

explicitly display the factor of β which we previously absorbed in the definition of the action. Our goal is to use the computer as a 'heat bath' at inverse temperature β.

The Monte Carlo simulation technique is quite old in statistical physics (Metropolis *et al.*, 1953). It provides the possibility of performing 'experiments' on virtual 'crystals' with interactions governed by an arbitrary Hamiltonian of choice. This in principle enables isolation of various dynamical features and their role in such phenomena as phase transitions. Furthermore, as the entire configuration is stored, any desired correlation function is in principle available. The technique converges well in both high and low temperature regimes and interpolates nicely in between. This latter point is of particular import to the particle physicist, who desires to relate the Wilson demonstration of confinement in strong coupling to the continuum field theory obtained in the weak coupling limit.

As with real experiments, Monte Carlo simulations have certain inherent sources of error. Statistical fluctuations are always present, and only decrease with the square root of the computing time. This can be a serious handicap if one is interested in some detailed parameter displaying fluctuations comparable to the signal. Then one must run a hundred times longer to merely reduce the errors to 10%. In addition, systematic effects may arise from the finite lattice size and spacing. For the four-dimensional systems considered here, the linear size of the lattice is necessarily quite limited, eight to ten sites on a side being typical. (At this writing, the largest lattice yet studied for a gauge theory had 16^4 sites; Bhanot and Rebbi, 1981.) Finally, a systematic error arises in determining when equilibrium has been reached; in particular, one must worry about being trapped in some metastable state.

Many of these systematic effects are readily amenable to further study. The lattice size is easily varied over a limited range and indeed observation of finite size effects can provide useful information on the states of the theory (Brower, Creutz and Nauenberg, 1982). Different initial conditions can test for thermal equilibrium; some possible starting states will be discussed later. Finite lattice spacing effects are of interest because they are tied to the renormalization of the bare coupling, as extensively discussed in chapters 12 and 13.

Regarding the computer as merely a heat bath immediately suggests the most intuitive Monte Carlo algorithm (Yang, 1963). We successively touch this heat bath to all the links in the lattice. A real thermal source in contact with a link would cause that variable to fluctuate thermally throughout the group manifold. When the source is removed, the link would be left in any of its allowed states with a probability given by the associated Boltzmann weight. For example, to mimic this process for the gauge group $Z_2 = \{1, -1\}$, one would begin by calculating the probability of the given link to be left in the state $+1$

$$P(1) = e^{-\beta S(1)}/(e^{-\beta S(1)} + e^{-\beta S(-1)}). \qquad (18.3)$$

Here $S(\pm 1)$ is the action evaluated with the link in question in the corresponding state and all other links held fixed at their current values. Note that if the action is local, that is if only nearby links are directly coupled, then this probability depends solely on these nearby links. The algorithm continues by asking the computer for a randomly selected number from a uniform distribution between zero and one. If $P(1)$ exceeds this number, the link is set to unity, otherwise it is set to -1. The entire procedure is then repeated on the next link and so forth until the entire lattice is covered. This represents one Monte Carlo iteration and generates the next state in a Markov chain of configurations.

These ideas are applicable to any group. The 'heat bath' algorithm replaces each group element with a new value selected randomly with a weighting given by the current exponentiated action. When applied to an ensemble of states, this gives a new ensemble which is closer to an equilibrium ensemble in a sense that we will shortly make precise.

When the group manifold is intricate, the above selection procedure for new group elements may be impractical or too time consuming to implement efficiently. For this reason computationally simpler algorithms are often used. These are also constructed to bring one closer to equilibrium, but may require more iterations to achieve the same convergence. If each iteration takes less computer time than a heat bath selection, there can be a net gain.

To design alternative procedures, we need a criterion for determining when an algorithm for randomly changing an ensemble of configurations will tend towards equilibrium. In general, each state in the Monte Carlo sequence results from a Markovian process applied to the previous configuration. Each stage in the algorithm is thus specified by a probability distribution $P(C', C)$ for taking configuration C into C'. An obvious necessary condition on P is that it leave an equilibrium ensemble in equilibrium. Thus the Boltzmann weights should be an eigenvector of P

$$e^{-\beta S(C)} = \sum_{C'} P(C, C') e^{-\beta S(C')}. \qquad (18.4)$$

Remarkably, if the algorithm has eventual access to any configuration, this is also a sufficient condition for any ensemble to ultimately approach the Boltzmann distribution of eq. (18.2).

To demonstrate this claim, we need a notion of 'distance' between ensembles. Suppose we have two ensembles E and E', each of many configurations. Denote the probability density for configuration C in E or E' by $p(C)$ or $p'(C)$, respectively. Then we define the distance between E and E' as the sum

$$\|E - E'\| = \sum_C |p(C) - p'(C)|, \tag{18.5}$$

where the sum is over all possible configurations. Now suppose that E' resulted from the application of a Monte Carlo algorithm satisfying eq. (18.4) to ensemble E. This means that

$$p'(C) = \sum_C P(C, C') p(C'). \tag{18.6}$$

As $P(C', C)$ is a probability, it satisfies

$$P(C', C) \geqslant 0, \tag{18.7}$$

$$\sum_C P(C', C) = 1. \tag{18.8}$$

We can now compare the distance of E' from equilibrium with the distance of E from equilibrium

$$\|E' - E_{eq}\| = \sum_C |\sum_{C'} P(C, C')(p(C') - p_{eq}(C'))|$$

$$\leqslant \sum_{C, C'} P(C, C') |p(C') - p_{eq}(C')| = \|E - E_{eq}\|. \tag{18.9}$$

We conclude that the algorithm reduces the distance of an ensemble from equilibrium. Note that if $P(C, C')$ never vanishes, this inequality is strict unless we are already in equilibrium.

To insure that an algorithm has the equilibrium distribution as an eigenvector, most algorithms in practice are based on products of steps each satisfying a condition of detailed balance

$$P(C', C) e^{-\beta S(C)} = P(C, C') e^{-\beta S(C')}. \tag{18.10}$$

Summing over the second index C' and using eq. (18.8) immediately gives the eigenvector eq. (18.4).

The detailed balance condition, which is sufficient but not necessary for the approach to equilibrium, far from uniquely specifies the matrix $P(C, C')$. The heat bath algorithm automatically satisfies the condition because $P(C, C')$ is independent of C' and proportional to the Boltzmann weight for C. Metropolis *et al.* (1953) used the detailed balance criterion to formulate another algorithm which, because of calculational simplicity, has become the most popular in practice. For the gauge theory, we begin with the selection of a trial U' as a tentative replacement for some link variable U. The test variable is selected with a distribution $P_T(U, U')$ which must be symmetric in U and U'

$$P_T(U, U') = P_T(U', U). \qquad (18.11)$$

Beyond this constraint, P_T is arbitrary and may be selected empirically to optimize convergence. Normally it is best if U' has a weighting towards the old value of U. Once U' is chosen, we evaluate the tentative new action $S(U')$ for comparison with its old value $S(U)$. If the action is lowered, that is if the new configuration has a larger Boltzmann weight, then this change is accepted. The detailed balance condition then determines the remainder of the algorithm: if the action is raised the change must be accepted with conditional probability $\exp(-\beta(S(U') - S(U)))$.

A simple way to implement this procedure is to obtain U' by multiplying U with a random group element from a table, where this table is itself of random elements with a convenient weighting towards the identity. The table should contain enough elements to generate for the group and should contain the inverse of each of its elements in order to satisfy eq. (18.11).

The Metropolis algorithm described above has an essential dependence on two parameters. The first is the weighting of the random changes towards the identity. This peaking should become more severe at low temperatures where large changes would be routinely rejected. A popular criterion for selecting this distribution is to make the acceptance probability at any step roughly 50%.

A second parameter in the algorithm is the number of trial changes attempted on any given link before going on to the next. In most statistical problems this is taken to be one; however, for the gauge theory the interaction is rather complicated and requires considerable arithmetic to evaluate. This means that it can be extremely beneficial to do as good a job as possible in selecting the stochastic changes. In terms of real computer time involved in reaching equilibrium, it is usually of value to test ten or more new elements, during which time the multiplication of neighboring elements appearing in the action need not be repeated. Note that as the number of tries, or 'hits' increases, the Metropolis algorithm approaches the heat bath. This is because repeating the procedure on one link will ultimately bring that link into thermal equilibrium with its temporarily fixed neighbors. This is what the heat bath does in one step. To determine an optimum number of hits, one can simply make a few trial runs on a small lattice to study the convergence in real time.

Although the Metropolis procedure brings an ensemble closer to equilibrium by less per iteration than the heat bath, it has the advantage of being extremely simple. The detailed form of the group measure is not needed; the algorithm automatically generates it with a random walk around the group. Furthermore, to change the form of the action or group is straightforward. Nevertheless, for groups with simple enough manifolds the heat bath algorithm may be rather elegantly implemented.

To illustrate some techniques for generating variables with a given weight, we will now discuss the heat bath generation of $SU(2)$ elements for the gauge theory with the Wilson action (Creutz, 1980b). First we need a source of random numbers uniformly distributed between zero and one. Such generators are standard in most high level computer languages, and we assume a good one has been provided (Knuth, 1969). The important point for our purposes is that calls to such a generator are extremely fast, comparable to a multiplication, and thus usually represent only a minor part of the time of a simulation.

Given a source of random numbers with such a uniform distribution, we can easily produce a random sequence with an arbitrary distribution. Suppose we have some positive function $f(x)$ on the unit interval and wish to generate points with a weighting proportional to f. For simplicity assume that f is bounded; if not, make a change of variables to make it so. Without loss of generality, we assume that $f(x)$ is bounded by unity. Using the given random number generator, we obtain a trial number for the first element of our weighted sequence. Calling this number x, we obtain a second random number and accept x if the new random variable is less than $f(x)$. This is repeated many times to form a sequence of accepted values of x. As the probability of accepting any x is proportional to $f(x)$, the sequence has the desired weighting.

This process will be inefficient if the function f is strongly peaked. In this case we may need to generate many points before one is accepted. If one knows approximately where the peak is and its form, one may be able to change variables to spread it out. This forms the basis for the following $SU(2)$ algorithm.

While working on a particular link (ij), we need consider only the contribution to the action coming from the six plaquettes containing that link. If we denote by $\tilde{U}_\alpha, \alpha = 1, \dots, 6$, the six products of three link variables which interact with the link in question, then the probability distribution for the new link variable under the heat bath algorithm is

$$\mathrm{d}p(U) \sim \mathrm{d}U \exp\left(\tfrac{1}{2}\beta \mathrm{Tr}\left(U \sum_{\alpha=1}^{6} \tilde{U}_\alpha\right)\right). \qquad (18.12)$$

For $SU(2)$ the trace is automatically real. In chapter 8 we parametrized $SU(2)$ as the surface of a four-dimensional sphere

$$SU(2) = \{a_0 + \mathrm{i}\mathbf{a}\cdot\sigma \,|\, a_0^2 + \mathbf{a}^2 = 1\}, \qquad (18.13)$$

where σ represents the Pauli matrices. The invariant group measure is uniform over this sphere

$$dU \sim d^4a\,\delta(a^2 - 1). \tag{18.14}$$

This representation shows the useful property that a sum over any number of $SU(2)$ elements is proportional to another element of the group. In particular, we have

$$\sum_{\alpha=1}^{6} \tilde{U}_\alpha = k\bar{U}, \tag{18.15}$$

where $\bar{U}$ is an element of $SU(2)$ and k is the determinant

$$k = \left| \sum_{\alpha=1}^{6} \tilde{U}_\alpha \right|^{\frac{1}{2}}. \tag{18.16}$$

The utility of this observation appears when we use the invariance of the group measure to absorb $\bar{U}$

$$dp(U\bar{U}^{-1}) \sim dU \exp\left(\tfrac{1}{2}\beta k \operatorname{Tr} U\right) \sim d^4a\,\delta(a^2 - 1)\exp(\beta k a_0). \tag{18.17}$$

Thus we have found the peak in the exponentiated action and rotated it to the identity. We have reduced the problem to generating points randomly on the surface of the unit sphere in four dimensions with an exponential weighting along the a_0 direction. Given an element U generated in this manner, we replace the link variable on the lattice with the product

$$U'_{ij} = U\bar{U}^{-1}. \tag{18.18}$$

To generate the appropriately weighted points on the sphere, we first do the integration over the magnitude of $\mathbf{a}$ with the delta function and obtain

$$dU \exp\left(\tfrac{1}{2}\beta k \operatorname{Tr} U\right) \sim \tfrac{1}{2}da_0 d\Omega\,(1 - a_0^2)^{\frac{1}{2}}\exp(\beta k a_0). \tag{18.19}$$

Here $d\Omega$ is the differential solid angle for the vector $\mathbf{a}$, which has length $(1 - a_0^2)^{\frac{1}{2}}$. We need the generate a_0 in the interval $(-1, 1)$ with probability density

$$dp(a_0) \sim (1 - a_0^2)^{\frac{1}{2}}\exp(\beta k a_0)\,da_0 \tag{18.20}$$

and the direction for $\mathbf{a}$ is totally random. For moderate to large β, the dominant peaking in eq. (18.20) comes from the exponential factor. This can be removed with a change of variables from a_0 to

$$z = \exp(\beta k a_0). \tag{18.21}$$

Equation (18.20) now becomes

$$dp(z) = dz\,(1 - \beta^{-2}k^{-2}\log^2 z)^{\frac{1}{2}}. \tag{18.22}$$

The generation of z can proceed as outlined earlier; with the random number generator a trial z is selected randomly in the allowed interval

$$e^{-2\beta k} \leqslant z \leqslant e^{+2\beta k} \tag{18.23}$$

and this is rejected with the probability given on the right hand side of eq. (18.22). Repeating this until a z is accepted, one reconstructs a_0 by taking a logarithm. The final step in the algorithm is to randomly select the direction for **a**. This can be done in a variety of ways; for example, one could generate a random point in the interior of a three-dimensional sphere and use its direction from the origin. Note that in the above discussion several tricks special to the group $SU(2)$ were used. To find corresponding tricks for a new group or even a new action can be tedious and thus most simulations in practice have turned to the Metropolis algorithm.

Monte Carlo computer programs tend to be physically rather short and straightforward. They begin with a set of nested loops over the various links. The selection of the new variables, such as outlined above, involves only a few rather simple operations. The multiple loops result in these steps being repeated on the order of a million times. The $SU(2)$ procedure is readily implementable so that it requires less than 200 microseconds per link on a *CDC* 7600 computer. The group $SU(3)$ with a reasonably optimized algorithm uses one to two milliseconds per link on the same machine. In both these cases, the majority of the time is spent multiplying the neighboring group elements. In practice it is usually computer time rather than storage which limits these programs. For $SU(2)$ it is convenient to store the four components of a_μ for each link, resulting in a relatively modest 160 000 numbers for a 10^4 site lattice.

We now turn to describe some simple Monte Carlo 'experiments'. An obvious first question involves the time required to reach equilibrium. When we are not operating near a phase transition this time can be remarkably short. In figure 18.1 we show the results of several runs with the heat bath algorithm on the group $SU(2)$. The coupling constant was set to the constant value

$$\beta = 4g_0^{-2} = 2.3 \qquad (18.24)$$

which was selected as representative of the slowest convergence occurring with this model. Runs are shown on four-dimensional lattices of from 4^4 to 10^4 sites. Each iteration represents one application of the heat bath algorithm to every lattice link; on the 10^4 lattice one such step represents 40 000 new $SU(2)$ elements. As a function of the number of iterations, we plot the average plaquette or expectation of the action

$$P = \langle 1 - \tfrac{1}{2} \mathrm{Tr}\, U_\square \rangle, \qquad (18.25)$$

discussed in chapter 9. For each size lattice, two different initial configurations were studied. The + symbols represent an ordered start, with all link matrices set to the identity. This ground state of the statistical system corresponds to beginning at zero temperature. In contrast, the crosses represent an initial configuration where each element was selected randomly, uniformly in the invariant measure, from the entire group. In this case we start at infinite temperature. Thus we approach equilibrium from opposite extremes. Note that in all cases convergence appears to be essentially complete after only 20 to 30 iterations. Thermal fluctuations, which must always be present, are quite apparent on the small lattices but become relatively small on the 10^4 site system.

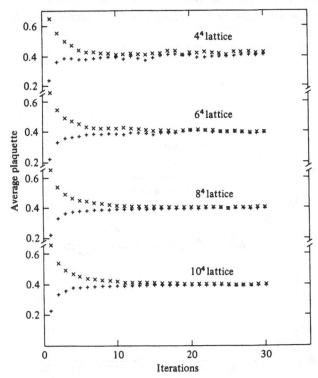

Fig. 18.1. Several Monte Carlo runs with the group $SU(2)$ (Creutz, 1980*b*).

The situation can be considerably less advantageous if a phase transition is nearby. In figure 18.2 we show the convergence of the $U(1)$ theory on a 6^4 lattice at $\beta = 1.0$. On an infinite lattice this system exhibits a second-order phase transition at $\beta = 1.012 \pm .005$ (Lautrup and Nauenberg, 1980*a*; DeGrand and Toussaint, 1980; Bhanot, 1981). In addition to the rather slow convergence when compared to the $SU(2)$ case, note the large fluctuations, characteristic of critical behavior.

The above runs illustrate the two simplest initial conditions, corresponding to zero and infinite temperature. In the case of a first-order transition such initial states can result in the lattice being caught in a metastable state. As in a real experiment, a random (ordered) lattice can be supercooled (superheated) substantially below (above) the transition temperature without settling in a reasonable time into the correct phase. To aid in approaching equilibrium one can add a 'seed' consisting of an ordered (disordered) piece of the lattice. This motivates a third interesting initial configuration consisting of a lattice which is half ordered and half disordered. For example, links from sites with fourth coordinate less than half the lattice size could be randomized and the remainder ordered. In

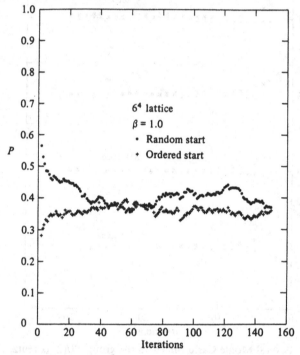

Fig. 18.2. Two Monte Carlo runs with the $U(1)$ model near its critical point.

figure 18.3 we show several Metropolis algorithm runs with such a start for the gauge group Z_2 on an 8^3 by 20 lattice (Creutz, Jacobs, and Rebbi, 1979b). Several values of temperature are selected near the transition point as calculated in chapter 16. The figure shows a linear drift characteristic of one phase 'dissolving' the other. The aimless drift very near the transition indicates that this method can rather accurately determine the temperature of the phase change. Indeed, this is analogous to putting some ice in water to determine its melting point.

The fact that convergence is fast away from phase transitions and slow near them suggests another type of experiment. Upon heating and then cooling the system through a range of temperatures, regions of slow convergence will appear as hysteresis effects. This provides a technique for rapidly locating regions of coupling for further study. In figure 18.4 we show the results of such thermal cycling on the $SU(2)$ model in four and five space-time dimensions and the $U(1) = SO(2)$ theory in four dimensions. Each point in this figure was obtained by running on the order of twenty iterations with the heat bath algorithm from either a hotter or cooler

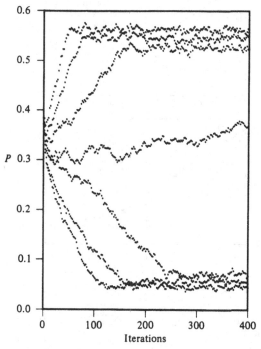

Fig. 18.3. Several Monte Carlo runs for the Z_2 model with mixed phase initial conditions. From the top down, these runs take β from 0.41 to 0.47 in steps of 0.01 (Creutz, Jacobs and Rebbi, 1979a).

configuration. As a check on normalizations, we also plot the lowest order strong and weak coupling results. Phase transitions are to be suspected in those regions where the heating and cooling cycles do not agree, as clearly observed for the five-dimensional $SU(2)$ and four-dimensional $U(1)$ models. Further analysis in the transition region suggests that the $U(1)$ transition is second order (Lautrup and Nauenberg, 1980a) and the five-dimensional $SU(2)$ transition is first order. As the latter fits the prediction of mean field theory, we conclude that $d = 5$ is close to $d = \infty$.

This is further supported by the fact that the $U(1)$ model in five dimensions also exhibits a first-order transition (Bhanot and Creutz, 1980).

The hysteresis seen in figure 18.4c may at first seem a bit surprising because the transition is believed to be second order and should have a continuous internal energy regarded as a function of the temperature.

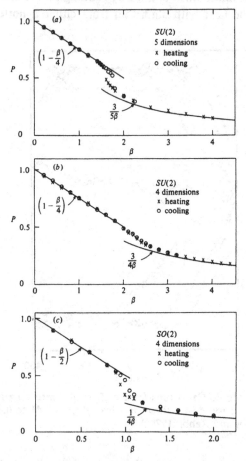

Fig. 18.4. Thermal cycles on several of the models (Creutz, 1979).

However, this thermal cycle was rather rapid, and, as figure 18.2 shows, a few tens of iterations are insufficient for relaxation of the energy near the critical point. Repeating this type of experiment at different cycle rates does indeed provide information on the nature of a transition. With a strong first order phase change, superheating and supercooling result in a hysteresis cycle which is reasonably stable in shape and relatively independent of the speed with which the system is heated or cooled. The closing of the cycle is predominantly determined by the temperatures at which metastable minima of the free energy disappear. In contrast, the cycle associated with a second-order transition tends to close continuously as the experiment becomes more adiabatic.

The four-dimensional $SU(2)$ model exhibits a sharp contrast to the other systems in figure 18.4. It displays no clear structure other than a fairly rapid crossover from strong to weak coupling behavior at β around two. Careful analysis of the specific heat in this region shows a definite peak but no signal of a real singularity (Lautrup and Nauenberg, 1980*b*). This result supports the desired absence of a phase transition in this non-Abelian system. In figure 11.1 we showed the internal energy of the $SU(3)$ model as obtained from Monte Carlo analysis. It exhibits a rapid crossover qualitatively similar to the $SU(2)$ case.

Problem

1. On your home computer, write a Monte Carlo program to simulate the one-dimensional Ising model. Calculate the internal energy as a function of temperature and compare with the exact result.

2. Devise a heat bath algorithm for the gauge group $U(1)$.

19

Monte Carlo simulation II: measuring observables

Regarding the lattice as merely an ultraviolet cutoff, we would like to use the Monte Carlo simulation technique for the calculation of some physical numbers characteristic of the continuum field theory. At the outset it is not clear how well this can be done with the rather limited lattices available. For believable results we must make the lattice spacing smaller than relevant hadronic scales and yet have the overall lattice larger than the scale of physics we are measuring. A lattice of order 10 sites in any given direction leaves little leeway in such an analysis. Furthermore, the renormalization group discussion of chapter 13 points out that we should expect an exponential dependence of the lattice spacing on coupling constant. At best only a very narrow range of coupling can be useful in extracting physical numbers.

To counteract this pessimism, we have the remarkable experimental fact that the scaling behavior predicted by asymptotic freedom appears in deep inelastic scattering experiments at the precociously low momentum transfers of order 2 GeV (Perkins, 1977). Thus our 10^4 site lattices may give interesting results for physics at energy scales down to a few hundred MeV, exactly where strong confinement forces should come into play. Thus we may hope to relate a few features of long- and short-distance quark dynamics.

We should attempt to measure a quantity which has a finite continuum limit; that is, we must extract a physical observable. The average plaquette, which dominated the above Monte Carlo discussion, is proportional to the expectation value of the action density and is expected at the perturbative level to have ultraviolet divergences. The simplest physical observable for extraction from a Monte Carlo analysis is K, the coefficient of the linear long-range interquark potential. This may be found by measuring large Wilson loops and looking for the area law falloff discussed in chapter 9. Measuring distances in lattice units, one actually determines the dimensionless combination a^2K as a function of the bare coupling g_0^2. If the linear potential survives the continuum limit, the weak coupling behavior of this combination should follow the prediction of the renormalization group as discussed in chapter 13.

$$a^2K = (K/\Lambda_0^2)(\gamma_0 g_0^2)^{(-\gamma_1/\gamma_0^2)} \exp(-1/(\gamma_0 g_0^2))(1+O(g_0^2)). \quad (19.1)$$

Conversely, verification of this behavior will provide strong evidence for the survival of the linear potential when the cutoff is removed.

In general the behavior of a Wilson loop can be quite complicated. In addition to the area law piece dominant for large loops, there should be perimeter dependence from the self energies of the quark sources and yet further corrections from perturbative gluon exchange across the loop. As the coupling is reduced and the continuum limit approached, the perimeter piece should diverge and dominate for any fixed size loop. To eliminate this distraction, it is convenient to consider ratios of loops with different areas but the same perimeter. In particular, define (Creutz, 1980c)

$$\chi(I,J) = -\ln\left(\frac{W(I,J)\,W(I-1,J-1)}{W(I,J-1)\,W(I-1,J)}\right), \quad (19.2)$$

where $W(I,J)$ denotes the expectation of a rectangular Wilson loop of lattice dimensions I by J. In these quantities any perimeter dependence or constant factors in the loops will cancel. Whenever the loops are dominated by an area law, $\chi(I,J)$ directly measures the string tension

$$\chi \to a^2K. \quad (19.3)$$

This occurs either when I and I are large or when the bare coupling is large. However, in the weak coupling limit gluon exchange should dominate and χ will have a perturbative expansion

$$\chi(I,J) = O(g_0^2). \quad (19.4)$$

For example the weak coupling expression for the one-by-one loop implies

$$\chi(1,1) = \left\{\begin{array}{ll} 3g_0^2/16+O(g_0^4), & SU(2) \\ g_0^2/3+O(g_0^4), & SU(3). \end{array}\right\} \quad (19.5)$$

Such a power behavior is in marked contrast to the essential singularity on the right hand side of eq. (19.26). To summarize, for strong coupling we expect all $\chi(I,J)$ to become the area law coefficient but as g_0^2 is reduced, smaller loops should give a χ deviating from the true a^2K. Thus the curves of $\chi(I,J)$ for all I and J should form an envelope along the curve of the string tension. For weak coupling this envelope should satisfy eq. (19.1).

In figure 19.1 we plot $\chi(I,I)$, for I up to four, as a function of g_0^{-2} for the gauge group $SU(2)$. At strong coupling the large loops have large relative errors but are consistent with χ approaching the values from smaller loops. The graph also indicates the strong coupling limit for the string tension

$$a^2K = \ln(g_0^2) + O(g_0^{-4}). \quad (19.6)$$

The weak coupling behavior of eq. (19.1) is shown as a band representing values of the parameter Λ_0 in the range

$$\Lambda_0 = (1.3 \pm 0.2) \times 10^{-2} K^{\frac{1}{2}}, \quad SU(2). \qquad (19.7)$$

This error is a purely subjective estimate.

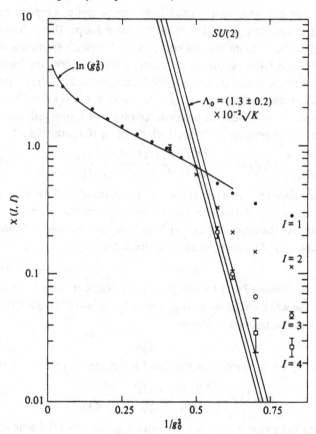

Fig. 19.1. Extracting the string tension for $SU(2)$ (Creutz, 1980c).

Figure 19.2 shows the same quantities for the gauge group $SU(3)$ on a 6^4 site lattice (Creutz and Moriarty, 1982b). On this smaller lattice, only loops up to 3-by-3 were used. For this theory the strong coupling expansion gives

$$a^2 K = \ln(3g_0^2) + O(g_0^{-2}). \qquad (19.8)$$

Note that the corrections for $SU(3)$ begin in a lower order of the strong coupling expansion than for the $SU(2)$ case in eq. (19.6). The weak coupling band for Λ_0 is now

$$\Lambda_0 = (6.0 \pm 1.0) \times 10^{-3} K^{\frac{1}{2}}, \quad SU(3). \qquad (19.9)$$

When first obtained, these small numbers were quite surprising, coming as they do from theories with no small dimensionless parameters. However, as discussed in chapter 13, the value of Λ_0 is strongly dependent on renormalization scheme. There we quoted the results of Hasenfratz and

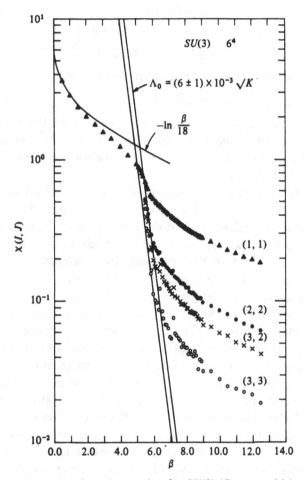

Fig. 19.2. Extracting the string tension for $SU(3)$ (Creutz and Moriarty, 1982b).

Hasenfratz, relating the lattice Λ_0 to the more conventional scale Λ_R defined by the three-point vertex in Feynman gauge and at a given scale in momentum space

$$\Lambda_R = \begin{cases} 57.5\Lambda_0, & SU(2) \\ 83.5\Lambda_0, & SU(3). \end{cases} \tag{19.10}$$

These large factors compensate the small numbers for Λ_0.

To compare these results with real experiments, we need some idea as to the expected value for K. This is provided by the string model (Goddard *et al.*, 1973), wherein a rotating string gives rise to a Regge trajectory of particle states. The slope of this trajectory in terms of the string tension is

$$\alpha' = (2\pi K)^{-1}. \tag{19.11}$$

Using the phenomenological $\alpha' = 1.0 \text{ GeV}^{-2}$, we find

$$K^{\frac{1}{2}} = 400 \text{ MeV} = (14 \text{ tons})^{\frac{1}{2}}. \tag{19.12}$$

Combining this with eq. (19.9) and (19.10), we conclude

$$\Lambda_R = 200 \pm 35 \text{ MeV}. \tag{19.13}$$

The current phenomenological value for this parameter is rather uncertain but consistent with this value. Such a direct comparison should be regarded with some caution, however, because the above calculation does not take account of virtual quark effects.

Despite its uncertainties, the above analysis is a rather remarkable, first principles calculation of a physical parameter relating opposite distance extremes. The scale Λ relates to short-distance scaling phenomena and K represents long-distance confinement effects. Their ratio is a number totally inaccessible to perturbative treatments. It characterizes the solution of a hopefully non-trivial four-dimensional field theory.

A second number of interest for Monte Carlo analysis is the mass gap or correlation length of the theory. This was discussed as a possible order parameter in chapter 9. In the pure gluon theory this is the mass of their lightest bound state, often referred to as a 'glueball.' In principle this quantity appears in a Yukawa law falloff of the correlation between two widely separated sources. Attempts to directly look for such correlations between plaquette operators have been plagued with statistical errors (Bhanot and Rebbi, 1981). Indeed, this correlation becomes swamped by the thermal fluctuations for a separation exceeding only a few lattice spacings. This problem can be circumvented with a combination of a variational method with the Monte Carlo analysis. For the plaquette–plaquette correlation at a short separation of only one or two sites, more than just exchange of the lightest state will be important. This means that a fit to a Yukawa law falloff will give an upper bound on the glueball mass. Using a linear combination of simple operators, for example loops of perimeter up to six links, and finding that combination that minimizes the falloff of the correlation with separation, one can improve the upper bound to a reliable estimate. Effectively, one is attempting to construct an operator which projects the desired state out of the spectrum. This analysis is usually further simplified by projecting out states of zero momentum. This is easily accomplished with a sum over translations transverse to the correlation distance.

Repeating this analysis for several values of coupling gives the functional dependence of the correlation length measured in lattice units. As with the string tension, one can check for the exponential dependence predicted by the renormalization group. The coefficient of this behavior gives the glueball mass in units of the lattice parameter Λ_0. The use of operators with certain discrete lattice symmetries readily generalizes the method to extract the mass of the lightest state with a given set of quantum numbers.

Several groups have developed this technique for the pure $SU(2)$ and $SU(3)$ theories (Berg, Billoire, and Rebbi, 1982; Berg and Billoire, 1982ab; Ishikawa, Schierholz and Teper, 1982). For the lightest state with $SU(3)$ these authors find

$$m/\Lambda_0 = 300\text{--}350. \tag{19.14}$$

In physical units this represents 700–1000 MeV, an experimentally intriguing value, although the effects of mixing with normal quark states are unknown. Going on to other quantum numbers, the above authors suggest extremely rich physics in the 1–2 GeV range.

We now come to a third physical parameter which is relatively easy to extract from the Monte Carlo analysis but more difficult to compare with real experiments. In chapter 3 we noted that a finite time length for the lattice permitted the study of finite physical temperatures in the physical quantum system. Thus using a four-dimensional lattice which is smaller in one direction than the others enables us to study the quantum statistical mechanics of pure non-Abelian gauge fields. Actually, the time dimension of the lattice may be varied in a combination of two ways, one by reducing the number of sites in that direction and the other by changing the timelike lattice spacing by means of a different coupling on timelike plaquettes, as used in the Hamiltonian discussion of chapter 15.

The interest in such finite temperature studies is the expectation of a real phase transition (Polyakov, 1978; Susskind, 1979). At low temperatures we should have the quark-confining vacuum with thermal fluctuations producing a dilute ideal gas of glueballs. At high temperatures, however, the vacuum can fill with a spaghetti of flux tubes. In such a pasta, an extra flux tube from an odd quark would quickly become lost. Thus we expect a transition to an unconfined phase in which quarks can wander freely away from each other.

We can regard our finite time system as representing the classical statistical mechanics of a three-dimensional slab of link variables. The deconfining transition corresponds to the spontaneous breaking of a global symmetry in this model. Consider a spacelike hypersurface passing between the sites of the slab, and consider multiplying each link variable that passes through this surface by an element from the center of the gauge group, for example -1 for $SU(2)$. Any plaquette must pass through the hypersurface an even number of times, equally in the two possible directions, and the extra factors will cancel. The action for the statistical system thus has a global symmetry under the center of the gauge group.

To monitor this symmetry, one can define Wilson loops with a net winding number in the timelike direction on the periodic lattice. The simplest such loop is just the trace of the product of all timelike links associated with a particular three-space position. In our toroidal geometry, such a loop is actually a straight line, a 'Wilson line'. These loops must each pass through the spacelike hypersurface an odd number of times and are thus not invariant under the global symmetry. We thus have an order parameter in the sense that a signal for the deconfining transition is the appearance of a spontaneous 'magnetization' with such loops.

A quarklike source on the lattice would produce a periodic world line along one of these loops. For an isolated quark, the product of link variables along this world line gives the gauge field interaction with the source. A vanishing expectation value for the loop is indicative of an infinite energy for an isolated quark in the confined phase. In contrast, a finite 'magnetization' represents the self energy of the quark in interaction with the gauge field soup.

To see that such a transition might well be expected, temporarily consider the extreme case of a one time-site lattice. The 'Wilson line' degenerates into the trace of the one timelike link at any given site. This link variable essentially degenerates into a spinlike variable. Indeed, for an Abelian group these 'spins' decouple from the spacelike loops and become a nearest-neighbor spin model in the three space dimensions. Such models in general have ferromagnetic transitions. For the non-Abelian case there remains a coupling the timelike and spacelike links, and we are left with a spin-gauge model with a global symmetry which can break spontaneously.

We have been discussing this deconfining transition in the pure glue theory without dynamical quarks. Remarkably, the plasma of flux tubes is sufficiently complicated that it can screen a source carrying a non-trivial representation of the gauge group center. This can never be accomplished with a finite number of gluons, each of which is in the adjoint representation and blind to the center. Note the contrast with an adjoint source, which would be screened in both phases. In the full theory with quark loops, quark pairs can always be 'popped' from the vacuum to screen any source. In this case it is unclear what to use for an order parameter, although presumably the existence of the deconfining transition is stable to the introduction of dynamical quarks.

Several groups have used these 'Wilson lines' to locate the critical temperatures for the pure $SU(2)$ and $SU(3)$ deconfining transitions (McLerran and Svetitsky, 1981; Kuti, Polonyi and Szlachanyi, 1981; Engels, Karsch, Satz and Montvay, 1981; Kajantie, Montonen and Pietarinen, 1981). Varying the bare coupling and the number of time sites, one can compare the dimensionless product of the lattice spacing and the critical temperature with the renormalization group prediction, in analogy to the string tension and mass gap analysis. For $SU(3)$ the observed transition temperature is

$$T_c/\Lambda_0 \approx 90 \qquad (19.15)$$

or in physical units

$$T_c \approx 200\,\text{MeV}. \qquad (19.16)$$

A peculiar feature of this number is its relative smallness in comparison to the glueball mass estimates. As we are considering the quarkless theory, the lightest states above the vacuum have energies large compared to eq. (19.16). This means that just below the transition temperature the vacuum is quite empty, only a low density of isolated glueballs are excited by thermal fluctuations.

An interesting unsettled question is the order of this transition (Svetitsky and Yaffe, 1982). For $SU(2)$ the symmetry being broken is Z_2 and presumably the transition is second order in analogy to the Ising model. However, for $SU(3)$ we have a Z_3 symmetry and the situation is less clear. Mean field theory for Z_3 systems typically predicts first-order transitions (problem 1 of chapter 14). As three dimensions is above the critical dimensionality for discrete symmetry breaking, this prediction must be considered seriously. Indeed, the simple Z_3 spin model, the three-state Potts (1952) model, does exhibit a first-order transition in three dimensions, although the latent heat is quite small (Blote and Swendsen, 1979). Current Monte Carlo studies of the $SU(3)$ deconfining transition are not yet able to determine its order. This question may have some relevance to the evolution of the very early universe.

Temperatures of the order in eq. (19.16) may be experimentally attainable for short times in heavy ion collisions. The relevance of the above calculations for this case is unclear for two reasons. First, such experiments entail a high quark density, and in the above discussion we considered the pure glue theory. Secondly, this temperature is above, although not by a large factor, a hypothetical maximum temperature of order 140 MeV where large numbers of pions would begin to be produced, consuming further kinetic energy forced into the system (Hagedorn, 1970). Indeed, other phase transitions related to pionic physics and/or chiral symmetry breaking may occur well before deconfinement is attained (Kogut *et al.*, 1982). In any case, temperatures in the few hundred MeV range promise rich physics for future experimental studies.

Up to this point our discussion of Monte Carlo simulation has avoided the question of Fermion fields. This is a rapidly evolving subject and therefore the remainder of this chapter is likely to soon be obsolete. The essential difficulty with including quarks in a numerical treatment is that the corresponding path integral is not an ordinary sum, but rather an intricate linear operation from the space of anticommuting variables into the complex numbers. Indeed, the exponentiated action is an operator and cannot be directly compared with real random numbers.

This problem can be immediately (foolishly?) circumvented by first integrating out the anticommuting variables analytically. As discussed in the chapter on fermionic integration, this gives a determinant when the action is quadratic in the anticommuting variables, as is usually the case in practice. This leaves us with an ordinary integral over the gauge fields, to which straightforward Monte Carlo methods are in principle applicable. The main difficulty with this approach is that the determinant is of an extremely large matrix, the number of rows being the product of the number of sites with the ranges of the spinor index, the internal gauge symmetry index, and the flavor index. For interesting sized systems, this is a many-thousand-dimensional matrix. As the time required to take a determinant of a matrix grows with the cube of its dimension, such direct calculations are prohibitively long. Furthermore, naively this determinant needs to be evaluated each time any gauge link is updated. Thus a simulation would seem to require evaluating an impossible determinant many thousands of times.

The actual situation is somewhat better because of various tricks. The fermionic matrix has an enormous number of zero elements. Because the interaction is local, no elements directly couple distant sites. Changes in a link variable alter only a small fraction of the remaining matrix elements. Considering a Metropolis *et al.* (1953) type of algorithm with a small step size, one can confine oneself to study small changes in a small part of the matrix. This still requires the inverse of the matrix, but as the gauge interaction will have stochastic errors anyway, hopefully one does not need the exact inverse. Approximate methods based on iterative schemes

(Weingarten and Petcher, 1981), Monte Carlo simulation with extra boson fields (Fucito *et al.* 1981; Scalapino and Sugar, 1981), and random walks through the matrix (Kuti, 1982) for finding the inverses of large matrices are under active investigation.

Despite these tricks, the fermionic problem is still extremely intensive in its demands on computer resources. An interesting approximation has been reasonably successful in approximately reproducing the hadronic spectrum (Hamber and Parisi, 1981; Weingarten, 1982; Marinari, Parisi and Rebbi, 1981). Instead of evaluating the determinant many times to allow it to feed back into the gauge field dynamics, this approximation considers the inversion of the fermionic matrix in a gauge field configuration obtained in a simulation of the pure quarkless theory. This determines how a quark would propagate in such a fixed background field. The basic approximation is the neglect of the feedback of the fermions on the gauge field. In perturbation theory, this amounts to the sum over all diagrams without any internal virtual quark loops. In a sense it represents the zero flavor limit.

Taking the expectation value of products of the propagators, one can study the propagation of bound state combinations with various meson or baryon quantum numbers. Neglecting internal loops in such systems amounts to considering only valence quarks and ignoring 'sea' quarks in the simple quark model. The experimental fact that a valence quark picture works fairly well suggests that the approximation may not be unreasonable. Internal loops are responsible for the ω splitting from the ρ meson, a relatively small effect. From a less optimistic point of view, neglecting virtual quark pairs neglects the decay of the ρ meson.

Mass estimates are obtained from the long-distance decrease of the meson and baryon propagators. The calculation begins with two parameters, the bare quark mass and the bare charge, which becomes related to the lattice spacing via the renormalization group. Thus two masses must be used to set these bare parameters. One is usually taken as the pion mass and the other either the Regge slope or the ρ mass. The most surprising result of these calculations is the ability to obtain a pion considerably lighter than the other hadrons with their typical scale of order one GeV. The approximation shows a signal of chiral symmetry breaking with the pion as a Goldstone boson. Note that one usually regards such a particle as a coherent excitation on a vacuum which is a condensate of elementary constituent pairs. To see such an effect while neglecting quark loops in the pion propagator is quite remarkable.

Several other predictions for observables should be available from this valence approximation. One can consider the valence quark propagation through three point vertices to obtain information on magnetic moments, form factors, and decay rates. The main missing feature of the procedure lies in states where strong mixing with pure glue states is important, as expected to be the case for the η and η' mesons.

As mentioned earlier, Monte Carlo work with fermions is rapidly evolving. Hopefully the above discussion of the difficulties with fermionic integration will soon become obsolete as the approaches and computer technology improve.

Problem

1. Derive the connection between the Regge slope and the string tension (eq. 19.11).

20

Beyond the Wilson action

The imposition of an ultraviolet cutoff is a highly non-unique procedure. Even in the framework of a lattice theory, innumerable variations are possible. Several decades of success with perturbative quantum electro-dynamics had led to the lore that the removal of any regulator yields the unique renormalized theory depending only on a small number of physical couplings and masses. Indeed, renormalizability is often regarded as a primary constraint on models for fundamental interactions.

On a non-perturbative level, however, little is rigorously known about even the existence of any four-dimensional theory, let alone its uniqueness. In some cases the theory may depend on even less parameters than suggested in perturbative analysis; for example, as discussed in chapter 13, Yang–Mills theories should undergo dimensional transmutation with dimensionless ratios being determined independent of any coupling constants.

The Monte Carlo technique provides an opportunity for non-perturbative exploration of cutoff dependence. Thus we can begin numerically to address these questions of the uniqueness of the continuum limit. In this chapter we discuss some of the simple variations of the Wilson scheme from this viewpoint of universality.

A simple alternative to the Wilson model places a vector field A_μ on the lattice sites and uses an action obtained by replacing derivatives in the continuum Yang–Mills Lagrangian with nearest-neighbor differences. This would be naively similar to the procedure followed in chapter 4 for scalar fields. This differs from the conventional lattice gauge theory in two important respects. First, the cutoff theory no longer has an exact local symmetry. This should not matter if the gauge breaking terms go away sufficiently rapidly in the continuum limit, but will complicate the renor-malization procedure. Second, the integral over gauges is no longer compact. The path integral will not be well-defined until gauge fixing is imposed. Because of its awkwardness, little work has been done with such a scheme, although Patrasciou, Seiler and Stametescu (1981) have done some preliminary Monte Carlo studies. They have not as yet seen the area law for large loops, but this is probably due to a renormalization of the bare charge making the linear potential appreciable only at extremely strong coupling.

Remaining closer in spirit to the Wilson formulation, Edgar (1982) considered replacing the plaquette with the two-by-one Wilson loop as the fundamental term in the action. In two space-time dimensions with the gauge group Z_2 this model is equivalent to the Ising model and therefore must have a phase transition, unlike the two-dimensional Wilson theory, which is trivial. The model possesses some extra global symmetries which can be broken; indeed, Edgar has seen a first-order phase transition in this '*fenêtre*' model with the gauge group $SU(2)$ in four dimensions. The moral of this is that the mere presence or absence of a phase transition is not a universal property of the gauge group. As we will see again later in this chapter, when the lattice spacing is not small, variations on the action can introduce new phenomena as lattice artifacts.

Drawing still closer to the Wilson theory, one can keep the action a class function of the group elements associated with the plaquettes, but change the detailed form of that function. We have already done that to some extent when we discussed duality and the Migdal–Kadanoff recursion relations, and we will pursue such generalizations further here. Manton (1980) presented a particularly simple alternative, taking for the action on a plaquette

$$S_{\square}(U) = \beta \, d^2(U, I), \qquad (20.1)$$

where d is the minimal distance in the group manifold between the element U and the identity I. The concept of a distance in the group manifold is formulated in terms of the metric tensor briefly mentioned in chapter 8. This metric is unique up to an overall normalization. In the case of $SU(2)$ the distance is simply

$$d(U_1, U_2) \propto \arccos\left(\tfrac{1}{2}\operatorname{Tr}\left(U_1 U_2^{-1}\right)\right). \qquad (20.2)$$

The Manton action is convenient for analytic work in the weak coupling limit. It is, however, singular for those elements with maximum distance from the identity, such as $-I$ for $SU(2)$. An amusing technical consequence of this singularity is that the transfer matrix is never positive definite (Grosse and Kuhnelt, 1981).

Another generalization, similar in spirit but different in detail from that of Manton, is the 'heat kernel' or generalized Villain (1975) action (Drouffe, 1978; Menotti and Onofri, 1981). This is based on the desire that the Boltzmann weight or exponentiated action

$$B(U_{\square}) = \exp\left(-S_{\square}(\beta, U_{\square})\right) \qquad (20.3)$$

should peak strongly near the identity element for weak coupling but should become uniform over the group for a simple strong coupling limit. This is reminiscent of expectation for the evolution of the temperature distribution in a piece of material shaped like the group manifold and initially possessing a spike in temperature at the identity. As time proceeds, the temperature spike should spread and eventually become uniformly distributed over the manifold. These ideas can be made mathematically precise using a group-theoretical generalization of the Laplacian to formulate a heat equation. Recall from chapter 8 the metric tensor

$$M_{ij} = \text{Tr}\,(g^{-1}(\partial_i g)\,g^{-1}(\partial_j g)), \qquad (20.4)$$

where the derivatives are with respect to the variables α_i which parameterize the group manifold. In terms of this, the invariant Laplace operator is given by the standard formula of differential geometry

$$\nabla^2 = \det(M)^{-\frac{1}{2}}(\partial/\partial\alpha_i)\det(M)^{\frac{1}{2}} M_{ij}^{-1}(\partial/\partial\alpha_j). \qquad (20.5)$$

We now define the heat equation

$$\nabla^2 K(t,g) = -(\mathrm{d}/\mathrm{d}t)\,K(t,g), \qquad (20.6)$$

where for convenience we have set the thermal diffusion coefficient to unity. For an initial condition we take

$$K(0,g) = \delta(g,I). \qquad (20.7)$$

The heat kernel action is directly identified with the solution of this equation at a time given by the coupling constant

$$e^{S_\Box(\beta,\,U)} = K(1/\beta, U). \qquad (20.8)$$

This action has the technical advantage over the Manton form of being smooth over the entire group manifold and giving rise to a positive definite transfer matrix.

Both the Manton and heat kernel actions have been subjected to Monte Carlo analysis (Lang, Rebbi, Salomonson and Skagerstam, 1981). The string tension was extracted as discussed in the last chapter. For comparison with the Wilson action results, the scheme dependence of the parameters must be calculated perturbatively. The results showed deviations of 20–40% from the theoretical expectations for their ratios, assuming that the physical string tension is universal. This should be regarded as the uncertainty due to the practical fact that the lattice spacing must be kept fairly large and therefore higher terms in the renormalization group function can be important.

Going on to another variant of the action, we note that an interesting change in the qualitative phase structure of the $SU(2)$ theory results from merely changing the trace of a plaquette to the corresponding trace in the adjoint representation (Greensite and Lautrup, 1981; Halliday and Schwimmer, 1981*a*). This amounts to working directly with the group $SO(3)$. In figure 20.1 we show a thermal cycle on this model with a 5^4 site lattice. Figure 20.2 shows the evolution of this system from ordered and

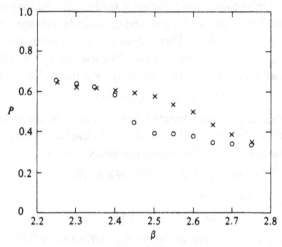

Fig. 20.1. A thermal cycle with $SO(3)$ lattice gauge theory on a 5^4 site lattice. The open circles represent heating; the crosses, cooling. (From Bhanot and Creutz, 1981.)

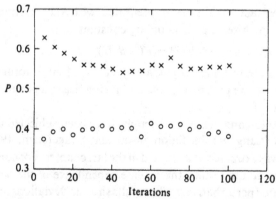

Fig. 20.2. Monte Carlo runs on $SO(3)$ lattice gauge theory at the transition temperature, $\beta = 2.5$. The open circles represent an ordered start, the crosses, random (Bhanot and Creutz, 1981).

disordered starts at the estimated transition temperature. These figures indicate a rather clear first-order transition.

As far as the classical limit is concerned, $SO(3)$ and $SU(2)$ Yang–Mills theories are identical. They only differ because of global properties which can come into play when quantum fluctuations bring plaquette operators far from the identity. The new transition is a lattice artifact which only shows up when the lattice spacing is not small. This is similar to the situation with the fenêtre action mentioned earlier.

One possible explanation of this $SO(3)$ transition is in terms of Z_2 monopole excitations. These arise because the $SO(3)$ representation of $SU(2)$ does not see the Z_2 center of the group. A plaquette variable near $-I$ in the group $SU(2)$ has the same energy as one near I. This can be used to define a Dirac string as a sequence of plaquettes near $-I$. Several closely related schemes for making this concept precise have been presented (Mack and Petkova, 1979; Tomboulis, 1981; Halliday and Schwimmer, 1981*b*; Brower, Kessler and Levine, 1982). We will follow Halliday and Schwimmer, who consider a slight modification of the theory. To make the action insensitive to the group center, they introduce a new set of variables $\{\sigma_\square\}$, each from the group $Z_2 = \{+1, -1\}$ and located on the lattice plaquettes. The new partition function is

$$Z = \sum_{\{\sigma_\square\}} \int (dU) \exp\left(\sum_\square \beta \sigma_\square \operatorname{Tr}(U_\square)\right). \tag{20.9}$$

As the action is linear in $\sigma_\square$, that part of the sum can be carried out to give

$$Z = \int (dU) \exp\left(\sum_\square S_\square(U_\square)\right), \tag{20.10}$$

where $$S_\square(U) = \log\left(2\cosh\left(\beta \operatorname{Tr} U\right)\right). \tag{20.11}$$

Being an even function of $\operatorname{Tr} U$, this quantity does not see the group center. Monte Carlo simulation (Halliday and Schwimmer, 1981*b*) has shown that this variation of the $SO(3)$ theory also has a first-order phase transition.

The quantity $\sigma_\square$ is essentially a Dirac string variable; when it is positive, $U_\square$ is weighted towards the identity, and when it is negative, $U_\square$ prefers to be near $-I$. The precise position of the Dirac string is unphysical because it can be moved around by absorbing factors of -1 into the link variables. However, in this process the ends of the string do not move; consequently, a natural definition of a monopole is to count the number of negative string variables entering any given three-dimensional cube and to say that a monopole is in that cube if this number is odd. On a four-dimensional lattice the monopoles will trace out world lines, and the strings sweep out world sheets. Halliday and Schwimmer measured the density of these monopole world lines in their simulation and found a sharp discontinuity at the transition temperature. The monopole density is not an order parameter in the sense of a magnetization for a spin system because thermal fluctuations prevent it from ever being exactly zero at any finite temperature. Nevertheless, it does provide a useful quantity to describe what is physically occurring at the transition.

The monopoles are easily supressed by giving them an ad hoc mass term. This motivates the more general partition function

$$Z = \sum_{\{\sigma_\square\}} \int (dU) \exp \left(\sum_\square \beta \sigma_\square \operatorname{Tr}(U_\square) + \lambda \sum_c \prod_{\square \epsilon c} \sigma_\square \right), \quad (20.12)$$

where the new sum in the exponent is over all three-dimensional cubes of the lattice. The presence of a monopole in any cube is now penalized by a factor of $e^{-2\lambda}$. As λ becomes large, the product of string variables over the surface of any cube must go to unity. An elementary exercise shows that once this has occurred there exists a set of Z_2 variables on the links such that any $\sigma_\square$ is the product of these around the given plaquette. In this event, all Z_2 factors are readily absorbed in the invariant $SU(2)$ measure and the theory goes over into the usual $SU(2)$ theory, which appears not to have any phase transitions. The limit $\beta \to 0$ in eq. (20.12) gives rise to a rather complicated looking Z_2 theory. However, under a duality transformation as discussed in chapter 16, this model turns into the usual four-dimensional Ising model with its second-order phase transition. Halliday and Schwimmer provided Monte Carlo evidence that as λ is increased, the $SO(3)$ transition moves to smaller β and eventually becomes the Ising transition. The place where the transition changes from first to second order is not known.

An alternative means for supressing monopoles is to add to the action of eq. (20.9) an effective potential for the variables $\sigma_\square$. Thus we could consider

$$Z = \sum_{\{\sigma_\square\}} \int (dU) \exp \left(\sum_\square \beta \sigma_\square \operatorname{Tr}(U) + \eta \sum_\square \sigma_\square \right). \quad (20.13)$$

As the parameter η goes to infinity, all $\sigma_\square$ are driven to unity and we again return to the pure $SU(2)$ theory. As $\sigma_\square$ is a Dirac string variable, the new term adds an effective energy per unit length to the strings. With η non-zero the strings become physical because moving them around will now change the total action in proportion to the total change in string length.

The action in eq. (20.13) is linear in $\sigma_\square$. These variables can be summed out to give an action dependent on the $U_\square$ only, as in eq. (20.10), but now

$$S_\square(U) = \log(2\cosh(\beta \operatorname{Tr} U + \eta)). \quad (20.14)$$

Unlike in eq. (20.11), this is no longer insensitive to the group center. Expanding this action in characters

$$S_\square(U) = \sum_R \beta_R \chi_R(U) \quad (20.15)$$

will give rise to terms with both integer and half-integer spin representations of $SU(2)$. Only the half-integer terms distinguish the group center. The action in eq. (20.14) has not been simulated directly, but it motivates a simpler form obtained by taking just the spin one-half and spin one terms in eq. (20.15) (Bhanot and Creutz, 1981).

$$S_\square(U) = \tfrac{1}{2}\beta \mathrm{Tr}\,(U) + \tfrac{1}{3}\beta_A \mathrm{Tr}_A(U). \qquad (20.16)$$

Here Tr_A denotes the trace or character in the adjoint or spin one representation. The factors in front of the couplings β and β_A are inserted for normalization convenience.

The theory defined by eq. (20.16) has several interesting limits. For vanishing β_A it reduces to the ordinary Wilson $SU(2)$ model, which we believe exhibits no phase transitions. In contrast, the limit of vanishing β gives the $SO(3)$ model, which we saw in figures 20.1 and 20.2 to have a first-order transition. The third interesting limit occurs as β_A goes to infinity. In this case all plaquettes are forced to lie in the center of the gauge group. This means that up to a gauge transformation all links are themselves driven to the center. Thus for $SU(2)$ the model becomes a Z_2 gauge theory with coupling β. As discussed in chapter 16, this model has a strong first-order phase transition at the self-dual point. At the outset, therefore, we know that the model of eq. (20.16) must have non-trivial phase structure, with two first-order lines entering the phase diagram.

Monte Carlo simulations have explored the evolution of these transitions into the two coupling plane (Bhanot and Creutz, 1981). The resulting phase diagram is shown in figure 20.3. Note that the $Z(2)$ and $SO(3)$ transitions are stable and meet at a triple point located at

$$(\beta, \beta_A) = (0.55 \pm 0.03, 2.34 \pm 0.03). \qquad (20.17)$$

A third first-order line extends from this point and aims toward the Wilson axis but terminates before reaching it at a critical point located at

$$(\beta, \beta_A) = (1.57 \pm 0.05, 0.78 \pm 0.05). \qquad (20.18)$$

This line points directly at the position of the peak in the specific heat of the ordinary $SU(2)$ model (Lautrup and Nauenberg, 1980b). That peak may be interpreted as a remnant of this transition, a shadow of its critical endpoint.

We can use this system to test the uniqueness of the continuum limit. The connection between the bare charge and the parameters is

$$g_0^{-2} = \beta/4 + 2\beta_A/3. \qquad (20.19)$$

A continuum limit requires taking g_0^2 to zero; however, this can be done along many paths in the (β, β_A) plane. Conventionally concentration is placed on the Wilson trajectory $\beta_A = 0$, $\beta \to \infty$. Along that line no singularities are encountered. Thus we have the usual claim that confinement, which is present in strong coupling, should persist into the weak coupling domain. However, an equally justified path would be, for example, $\beta = \beta_A \to \infty$. In this case we cross a first-order transition. Because one can continue around it in our larger coupling constant space, the transition is not deconfining and is simply an artifact of the lattice action.

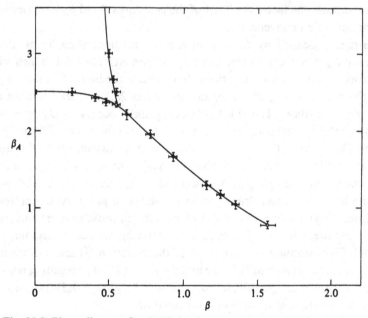

Fig. 20.3. Phase diagram for $SU(2)$ lattice gauge theory with fundamental and adjoint couplings (Bhanot and Creutz, 1981).

To test whether physical observables are indeed independent of direction in this plane, we can consider Wilson loops in the weak coupling regime. The loop by itself is not an observable because of self-energy divergences (Dotsenko and Vergeles, 1980; also recall problem 4 of chapter 6). These divergences should cancel in ratios of loops with the same perimeters and numbers of sharp corners. This leads us to consider the ratios

$$R(I, J, K, L) = \frac{W(I, J)\, W(K, L)}{W(I, L)\, W(J, K)}, \tag{20.20}$$

where $W(I, J)$ denotes the rectangular Wilson loop of dimensions I-by-J in lattice units. Wishing to compare points which give similar physics, we can consider for each value of β_A the value of β for which some R ratio has a particular value. In figure 20.4 we show points from Monte Carlo simulation for $R(2,2,3,3)$ having the values 0.87 and 0.93. The dashed lines in the figure represent constant bare charge from eq. (20.19). This particular simulation was performed with a 120-element subgroup approximating $SU(2)$. This is a good approximation where we are working, but does give rise to an extra transition to a highly ordered state at large

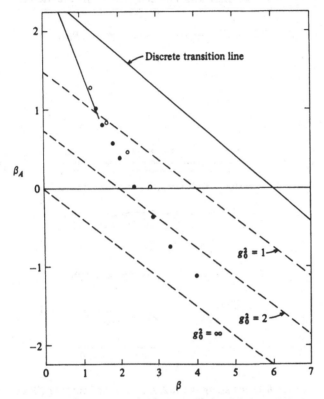

Fig. 20.4. Points of constant 'physics' as obtained from $R(2,2,3,3) = 0.87$ (solid circles) and 0.93 (open circles) (Bhanot and Creutz, 1981).

values of inverse coupling. The location of this 'discreteness' transition line is also indicated in figure 20.4.

If physics is indeed similar at all points along one of these contours of constant R ratio, then it should not matter which ratio we chose. In figure 20.5 we show several such ratios as functions of β_A along the $R(2,2,3,3) = 0.87$ contour. The comparison is quite good considering that finite cutoff corrections are ignored. Note that in this comparison the bare charge is far from being a constant. Along the 0.87 contour of 'constant physics', g_0^2 varied from less than unity to nearly 4 in the measured region. Such variation is permissible and perhaps even expected since the bare charge is unobservable and should depend on the cutoff prescription. The dependence can be characterized by a β_A dependent renormalization scale $\Lambda_0(\beta_A)$. The expected dependence of the renormalization scale on the new coupling β_A is calculable in perturbation theory (Gonzales-Arroyo and Korthals-Altes, 1982; Bhanot and Dashen, 1982). In the vicinity of the

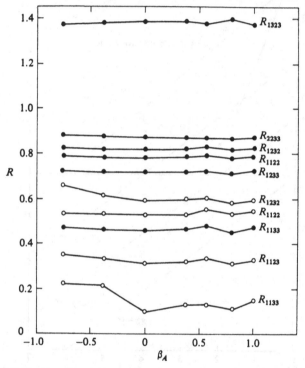

Fig. 20.5. Various R ratios along the $R(2, 2, 3, 3) = 0.87$ contour (Bhanot and Creutz, 1981).

Wilson action, that is when $|\beta_A a| < 0.5$, the prediction works reasonably well. However, as we approach the critical endpoint at positive β_A, large deviations from the points in figure 20.4 are found. This indicates that important additional physics is affecting the Monte Carlo results. We are seeing lattice artifacts near the new critical point. For negative β_A where the bare coupling becomes large, the agreement with the perturbative result is again poor. This can presumably be understood because of higher terms in the renormalization group function coming into play as the coupling increases (Grossman and Samuel, 1983).

This analysis indicates the privileged role played by the Wilson action. It appears to lie in a middle region where the scaling of asymptotic freedom appears on the modest lattices available to Monte Carlo simulation.

As the parameter β_A increases relative to β, the extremum of the action at $U = -I$ changes from a maximum to a minimum. This occurs along the line

$$\beta_A = 3\beta/8. \tag{20.21}$$

Finally, along the β_A axis the two minima are degenerate. Note that the critical endpoint lies slightly above the line in eq. (20.21). Bhanot (1982) has studied a similar two-coupling $SU(3)$ theory and finds a critical endpoint near the appearance of new minima of the plaquette action for group elements lying in the group center. As the n of $SU(n)$ increases beyond four, those elements of the group center near the identity become minima of the action even for the conventional Wilson action (Bachas and Dashen, 1982). This observation correlates well with the Monte Carlo results that the Wilson $SU(4)$, $SU(5)$ and $SU(6)$ theories all display first-order phase transitions (Creutz, 1981b; Moriarty, 1981; Creutz and Moriarty, 1982a). Presumably a negative β_A removes the extraneous action minima and will permit continuation around these transitions, which would therefore not be deconfining.

In the last few chapters we have seen that Monte Carlo simulation indeed provides a powerful tool. The technique not only permits calculation of observables, but also opens a way to investigate questions of existence and uniqueness. These investigations of the solutions of non-trivial quantum field theories indicate that we are truly at an exciting time in the development of elementary particle physics.

Problems

1. Show that if the product of the $\sigma_\square$ variables in eq. 12 is unity for every three-dimensional cube, then these parameters can be written as the product of Z_2 variables on the links surrounding the corresponding plaquettes.

2. Verify the assertion that the $\beta_A \to \infty$ limit of the theory defined by eq. (20.16) is indeed a Z_2 gauge theory.

3. Consider a three-parameter generalized $SU(2)$–$SO(3)$ action with both the λ term of eq. (20.12) and the η term of eq. (20.13). Discuss the various two parameter limits of this model.

References

Abbott, L. F. and E. Farhi, 1981a, *Phys. Lett.* **101B**, 69.
Abbott, L. F. and E. Farhi, 1981b, *Nucl. Phys.* **B189**, 547.
Adler, S. L., 1969, *Phys. Rev.* **117**, 2426.
Aharonov, Y. and D. Bohm, 1959, *Phys. Rev.* **115**, 485.
Ashmore, J. F., 1972, *Nuovo Cim. Lett.* **4**, 289.
Bachas, C. and R. Dashen, 1982, *Nucl. Phys.* **B210**, 583.
Baker, G. A. Jr. and J. Kincaid, 1979, *Phys. Rev. Lett.* **42**, 1431.
Banks, T., S. Raby, L. Susskind, J. Kogut, D. R. T. Jones, P. N. Scharbach
   and D. K. Sinclair, 1977, *Phys. Rev.* **D15**, 1111.
Banks, T., S. Yankielowicz and A. Schwimmer, 1980, *Phys. Lett.* **96B**, 67.
Bars, I., 1976, *Phys. Rev. Lett.* **36**, 1521.
Bars, I., 1981, *Physica Scripta*, **23**, 983.
Balian, R., J. M. Drouffe and C. Itzykson, 1974, *Phys. Rev.* **D10**, 3376.
Balian, R., J. M. Drouffe and C. Itzykson, 1975a, *Phys. Rev.* **D11**, 2098.
Balian, R., J. M. Drouffe and C. Itzykson, 1975b, *Phys. Rev.* **D11**, 2104.
Becher, P. and H. Joos, 1982, *Zeit. Phys.* **C15**, 343.
Beg, M. A. B. and H. Ruegg, 1965, *J. Math. Phys.* **6**, 677.
Bell, J. and R. Jackiw, 1969, *Nuovo Cimento*, **60A**, 47.
Bender, C. M., F. Cooper, G. S. Guralnik, R. Roskies and D. H. Sharp, 1981,
   Phys. Rev. **D23**, 2976.
Berg, B. and A. Billoire, 1982a, *Phys. Lett.* **113B**, 65.
Berg, B. and A. Billoire, 1982b, *Phys. Lett.* **114B**, 324.
Berg, B., A. Billoire and C. Rebbi, 1982, *Ann. Phys.* **142**, 185.
Bhanot, G., 1981, *Phys. Rev.* **D24**, 461.
Bhanot, G., 1982, *Phys. Lett.* **108B**, 337.
Bhanot, G. and M. Creutz, 1980, *Phys. Rev.* **D21**, 2892.
Bhanot, G. and M. Creutz, 1981, *Phys. Rev.* **D24**, 3212.
Bhanot, G. and R. Dashen, 1982, *Phys. Lett.* **113B**, 299
Bhanot, G. and C. Rebbi, 1981, *Nucl. Phys.* **B180**, 469.
Blairon, J. M., R. Brout, F. Englert and J. Greensite, 1981, *Nucl. Phys.* **B180**
   [FS2], 439.
Blote, H. and R. Swendsen, 1979, *Phys. Rev. Lett.* **43**, 799.
Bollini, C. G. and J. J. Giambiagi, 1972, *Nuovo Cim.* **12B**, 20.
Brezin, E., J.-C. Le Guillou and J. Zinn-Justin, 1977a, *Phys. Rev.* **D15**, 1544.
Brezin, E., J.-C. Le Guillou and J. Zinn-Justin, 1977b, *Phys. Rev.* **D15**, 1558.
Brower, R. C., M. Creutz and M. Nauenberg, 1982, *Nucl. Phys.* **B210** [FS6],
   133.
Brower, R. C., D. A. Kessler and H. Levine, 1982, *Nucl. Phys.* **B205**, 77.
Brower, R. and M. Nauenberg, 1980, *Nucl. Phys.*, **B180**, 221.
Cardy, J., 1980, *J. Phys.* **A13**, 1507.
Caswell, W. E., 1974, *Phys. Rev. Lett.* **33**, 244.

Chodos, A. and J. B. Healy, 1977, *Phys. Rev.* **D16**, 387.

Chodos, A., R. L. Jaffe, K. Johnson, C. B. Thorn and V. F. Weisskopf, 1975, *Phys. Rev.* **D12**, 2060.

Coleman, S., 1973, *Comm. Math. Phys.* **31**, 259.

Coleman, S. and E. Weinberg, 1973, *Phys. Rev.* **D7**, 1888.

Coleman, S. and E. Witten, 1980, *Phys. Rev. Lett.* **45**, 100.

Collins, P. D. and E. J. Squires, 1968, *SP Physics*, **45**, 1.

Creutz, M., 1977, *Phys. Rev.* **D15**, 1128.

Creutz, M., 1978*a*, *Rev. Mod. Phys.* **50**, 561.

Creutz, M., 1978*b*, *J. Math. Phys.* **19**, 2043.

Creutz, M., 1979, *Phys. Rev. Lett.* **43**, 553.

Creutz, M., 1980*a*, *Phys. Rev.* **D21**, 1006.

Creutz, M., 1980*b*, *Phys. Rev.* **D21**, 2308.

Creutz, M., 1980*c*, *Phys. Rev. Lett.* **45**, 313.

Creutz, M., 1981*a*, *Phys. Rev.* **D23**, 1815.

Creutz, M., 1981*b*, *Phys. Rev. Lett.* **46**, 1441.

Creutz, M. and B. Freedman, 1981, *Ann. Phys.* (N.Y.) **132**, 427.

Creutz, M., L. Jacobs and C. Rebbi, 1979*a*, *Phys. Rev. Lett.* **42**, 1390.

Creutz, M., L. Jacobs and C. Rebbi, 1979*b*, *Phys. Rev.* **D20**, 1915.

Creutz, M. and K. J. M. Moriarty, 1982*a*, *Phys. Rev.* **D25**, 1724.

Creutz, M. and K. J. M. Moriarty, 1982*b*, *Phys. Rev.* **D26**, 2166.

Creutz, M. and T. N. Tudron, 1978, *Phys. Rev.* **D17**, 2619.

Dashen, R. and D. Gross, 1981, *Phys. Rev.* **D23**, 2340.

DeGrand, T. and D. Toussaint, 1980, *Phys. Rev.* **D22**, 2478.

Dirac, P. A. M., 1933, *Physik. Zeits. Sowjetunion* **3**, 64.

Dirac, P. A. M., 1945, *Rev. Mod. Phys.* **17**, 195.

Dotsenko, V. S. and S. N. Vergeles, 1980, *Nucl. Phys.* **B169**, 527.

Drell, S. D., M. Weinstein and S. Yankielowicz, 1976, *Phys. Rev.* **D14**, 1627.

Drouffe, J. M., 1978, *Phys. Rev.* **D18**, 1174.

Drouffe, J. M., 1981, *Phys. Lett.* **105B**, 46.

Dyson, F. J., 1952, *Phys. Rev.* **85**, 631.

Edgar, R. C., 1982, *Nucl. Phys.* **B200** [FS4], 345.

Eichten, E., K. Gottfried, T. Kinoshita, K. D. Lane and T.-M. Yan, 1980, *Phys. Rev.* **D21**, 203.

Elitzur, S., 1975, *Phys. Rev.* **D12**, 3978.

Elitzur, S., R. B. Pearson and J. Shigemitsu, 1979, *Phys. Rev.* **D19**, 3698.

Engels, J., F. Karsch, H. Satz and I. Montvay, 1981, *Phys. Lett.* **101B**, 89.

Fadeev, L. and V. Popov, 1967, *Phys. Lett.* **25B**, 29.

Feynman, R. P., 1948, *Rev. Mod. Phys.* **20**, 367.

Fisher, M. E. and D. S. Gaunt, 1964, *Phys. Rev.* **133**, A224.

Flyvbjerg, H., B. Lautrup and J. B. Zuber, 1982, *Phys. Lett.* **110B**, 279.

Fradkin, E. and S. Shenker, 1979, *Phys. Rev.* **D19**, 3682.

Freedman, B., P. Smolensky and D. Weingarten, 1982, *Phys. Lett.* **113B**, 481.

Frohlich, J., G. Morchio and F. Strocchi, 1981, *Nucl. Phys.* **B190** [FS3], 553.

Fucito, F., E. Marinari, G. Parisi and C. Rebbi, 1981, *Nucl. Phys.* **B180** [FS2], 369.

Gell-Mann, M. and F. Low, 1954, *Phys. Rev.* **95**, 1300.

Gell-Mann, M. and Y. Ne'eman, 1964, *The Eightfold Way* (Benjamin, N.Y.).

Goddard, P., J. Goldstone, C. Rebbi and C. B. Thorn, 1973, *Nucl. Phys.* **B56**, 109.

Goldstone, J., 1961, *Nuovo Cimento* **19**, 15.

Quarks, gluons and lattices

Gonzales-Arroyo, A. and C. P. Korthals-Altes, 1982, *Nucl. Phys.* **B205**, 46.
Greensite, J. and B. Lautrup, 1981, *Phys. Rev. Lett.* **47**, 9.
Gross, D. and F. Wilczek, 1973a, *Phys. Rev. Lett.* **30**, 1343.
Gross, D. and F. Wilczek, 1973b, *Phys. Rev.* **D8**, 3633.
Grosse, H. and H. Kuhnelt, 1981, *Nucl. Phys.* **B205**, 273.
Grossman, B. and S. Samuel, 1982, *Phys. Lett* **120B**, 383.
Guth, A. H., 1980, *Phys. Rev.* **D21**, 2291.
Hagedorn, R., 1970, *Nucl. Phys.* **B24**, 93.
Halliday, I. G. and A. Schwimmer, 1981a, *Phys. Lett.* **101B**, 327.
Halliday, I. G. and A. Schwimmer, 1981b, *Phys. Lett.* **102B**, 337.
Hamber, H. and G. Parisi, 1981, *Phys. Rev. Lett.* **47**, 1792.
Hasenfratz, A. and P. Hasenfratz, 1980, *Phys. Lett.* **93B**, 165.
Hasenfratz, A. and P. Hasenfratz, 1981, *Nucl. Phys.* **B193**, 210.
Higgs, P., 1964, *Phys. Rev. Lett.* **13**, 508.
Horn, D., M. Weinstein and S. Yankielowicz, 1979, *Phys. Rev.* **D19**, 3715.
Huebner, R. P. and J. R. Clem, 1974, *Rev. Mod. Phys.* **46**, 409.
Ishikawa, K., G. Schierholz and M. Teper, 1982, *Phys. Lett.* **110B**, 399.
Jones, D. R. T., 1974, *Nucl. Phys.* **B75**, 531.
Kadanoff, L. P., 1976, *Ann. Phys.* **100**, 359.
Kadanoff, L. P., 1977, *Rev. Mod. Phys.* **49**, 267.
Kajantie, K., C. Montonen and E. Pietarinen, 1981, *Zeit. Phys.* **C9**, 253.
Karsten, L. and J. Smit, 1981, *Nucl. Phys.* **B183**, 103.
Kawai, H., R. Nakayama and K. Seo, 1981, *Nucl. Phys.* **B189**, 40.
Kerler, W., 1981a, *Phys. Rev.* **D23**, 2384.
Kerler, W., 1981b, *Phys. Rev.* **D24**, 1595.
Kluberg-Stern, H., A. Morel, O. Napoli and B. Petersson, 1981, *Nucl. Phys.* **B190** [FS3], 504.
Knuth, D. E., 1969, *The Art of Computer Programming*, vol. II. Addison-Wesley.
Kogut, J. B., 1979, *Rev. Mod. Phys.* **51**, 659.
Kogut, J. B., R. B. Pearson, J. Shigemitsu and D. K. Sinclair, 1980, *Phys. Rev.* **D22**, 2447.
Kogut, J. B., M. Stone, H. Wyld, J. Shigemitsu, S. Shenker and D. K. Sinclair, 1982, *Phys. Rev. Lett.* **48**, 1140.
Kogut, J. and L. Susskind, 1974, *Phys. Rev.* **D9**, 3501.
Kogut, J. and L. Susskind, 1975, *Phys. Rev.* **D11**, 395.
Kogut, J. and K. Wilson, 1974, *Phys. Reports* **12**, 75.
Korthals-Altes, C. P., 1978, *Nucl. Phys.* **B142**, 315.
Kosterlitz, J. M. and D. J. Thouless, 1973, *J. Phys.* **C6**, 1181.
Kramers, H. A. and G. H. Wannier, 1941, *Phys. Rev.* **60**, 252.
Kuti, J., 1982, *Phys. Rev. Lett.* **49**, 183.
Kuti, J., J. Polonyi and K. Szlachanyi, 1981, *Phys. Lett.* **98B**, 199.
Lang, C. B., C. Rebbi, P. Salomonson and B. S. Skagerstam, 1981, *Phys. Lett.* **101B**, 173.
Lang, C. B., C. Rebbi, P. Salomonson and B. S. Skagerstam, 1982, *Phys. Rev.* **D26**, 2028.
LaRue, G. S., J. D. Phillips and W. M. Fairbank, 1981, *Phys. Rev. Lett.* **46**, 967.
Lautrup, B. and M. Nauenberg, 1980a, *Phys. Lett.* **95B**, 63.
Lautrup, B. and M. Nauenberg, 1980b, *Phys. Rev. Lett.* **45**, 1755.
Lipatov, L. N., 1977, *Zh. Eksp. Teor. Fiz. Pis'ma Red*, **25**, 116.

Mack, G. and V. B. Petkova, 1979, *Ann. Phys.* **123**, 442.
Mandelstam, S., 1962, *Ann. Phys.* **19**, 1.
Manton, N. S., 1980, *Phys. Lett.* **96B**, 328.
Marinari, E., G. Parisi and C. Rebbi, 1981, *Phys. Rev. Lett.* **47**, 1795.
Matthews, P. T. and A. Salam, 1954, *Nuovo Cim.* **12**, 563.
McLerran, L. and B. Svetitsky, 1981, *Phys. Lett.* **98B**, 195.
Meissner, W. H. and R. Ochsenfeld, 1933, *Naturwissenschaften*, **21**, 787.
Menotti, P. and E. Onofri, 1981, *Nucl. Phys.* **B190**, 288.
Mermin, N. D. and H. Wagner, 1966, *Phys. Rev. Lett.* **22**, 1133.
Metropolis, N., A. W. Rosenbluth, M. N. Rosenbluth, A. H. Teller and E. Teller, 1953, *J. Chem. Phys.* **21**, 1087.
Migdal, A. A., 1975a, *Zh. Eksp. Teor. Fiz.* **69**, 810 [*Sov. Phys. Jetp.* **42**, 413].
Migdal, A. A., 1975b, *Zh. Eksp. Teor. Fiz.* **69**, 1457 [*Sov. Phys. Jetp.* **42**, 743].
Moriarty, K. J. M., 1981, *Phys. Lett.* **106B**, 130.
Moriarty, K. J. M. and E. Pietarinen, 1982, *Phys. Lett.* **112B**, 233.
Munster, G., 1980, *Nucl Phys.* **B190** (FS3), 439; Errata: **B200**, 536 and **B205**, 648.
Munster, G. and P. Weisz, 1980, *Phys. Lett.* **96B**, 119.
Nambu, Y. and G. Jona-Lasinio, 1961, *Phys. Rev.* **124**, 246.
Nielsen, H. B. and M. Ninomiya, 1981a, *Nucl. Phys.* **B185**, 20.
Nielsen, H. B. and M. Ninomiya, 1981b, *Nucl. Phys.* **B193**, 173.
Osterwalder, K. and E. Seiler, 1978, *Ann. Phys.* **110**, 440.
Patrasciou, A., E. Seiler and I. O. Stametescu, 1981, *Phys. Lett.* **107B**, 364.
Pauli, W. and F. Villars, 1949, *Rev. Mod. Phys.* **21**, 434.
Peierls, R. E., 1935, *Ann. Inst. Henri Poincaré*, **5**, 177.
Perkins, D. H., 1977, *Rep. Prog. Phys.* **40**, 409.
Petermann and Stueckelberg, 1953, *Helv. Phys. Acta* **26**, 499.
Politzer, H. D., 1973, *Phys. Rev. Lett.* **30**, 1346.
Polyakov, A. M., 1975, *Phys. Lett.* **59B**, 79.
Polyakov, A. M., 1978, *Phys. Lett.* **72B**, 477.
Potts, R. B., 1952, *Proc. Cambridge Philos. Soc.* **48**, 106.
Rabin, J., 1982, *Nucl. Phys.* **B201**, 315.
Savit, R., 1980, *Rev. Mod. Phys.* **52**, 453.
Scalapino, D. J. and R. Sugar, 1981, *Phys. Rev. Lett.* **46**, 519.
Shigemitsu, J. and J. B. Kogut, 1981, *Nucl. Phys.* **B190** [FS3], 365.
Susskind, L., 1979, *Phys. Rev.* **D20**, 2610.
Svetitsky, B., S. D. Drell, H. R. Quinn and M. Weinstein, 1980, *Phys. Rev.* **D22**, 490.
Svetitsky, B. and L. Yaffe, 1982, *Phys. Rev.* **D26**, 963.
t'Hooft, G., 1974, *Nucl. Phys.* **B72**, 461.
t'Hooft, G., 1980, *Cargese Summer Institute Lectures (1979)*, Plenum Press, New York.
t'Hooft, G. and M. Veltman, 1972, *Nucl. Phys.* **B44**, 189.
Tomboulis, E., 1981, *Phys. Rev.* **D23**, 2371.
Ukawa, A., P. Windey and A. H. Guth, 1980, *Phys. Rev.* **D21**, 1013.
Villain, J., 1975, *J. Phys. (Paris)*, **36**, 581.
Wegner, F. J., 1971, *J. Math. Phys.* **12**, 2259.
Weinberg, S., 1965, *Phys. Rev.* **138**, B988.
Weingarten, D., 1980, *Phys. Lett.* **90B**, 280.
Weingarten, D., 1982, *Phys. Lett.* **109B**, 57.
Weingarten, D. and D. Petcher, 1981, *Phys. Lett.* **99B**, 333.

Weisz, P., 1981. *Phys. Lett.* **100B**, 331.
Wilson, K., 1971*a*, *Phys. Rev.* **B4**, 3174.
Wilson, K., 1971*b*, *Phys. Rev.* **B4**, 3184.
Wilson, K., 1974, *Phys. Rev.* **D10**, 2445.
Wilson, K., 1975, *Phys. Reports*, **23**, 331.
Wilson, K., 1977, In *New Phenomena in Subnuclear Physics*, Edited by A. Zichichi (Plenum Press, N.Y.).
Yang, C.-P., 1963, *Proc. of Symposia in Applied Mathematics*, Vol. XV (Amer. Math. Soc., Providence, R.I., 1963), 351.
Yang, C. N., 1975, *Phys. Rev. Lett.* **33**, 445.
Yang, C. N. and R. Mills, 1954, *Phys. Rev.* **96**, 191.
Yoneya, T., 1978, *Nucl. Phys.* **B144**, 195.

Index

adjoint representation, 31, 33, 62, 105, 146, 154–7
Aharonov–Bohm experiment, 33
anomaly, 21, 26, 76
anticommuting numbers, 20–8, 148–50
antiperiodic boundary conditions, 28
area law, 54–5, 62–7, 82, 140–3
asymptotic freedom, 6, 88–93, 124, 161
asymptotic series, 17
axial gauge, 58

background field, 91
bag model, 3
bare coupling, 17, 29, 81–93, 102, 128
bare mass, 37, 81, 92, 149
baryon, 3, 74–6, 149
Boltzmann weight, 60, 109–15, 127–31, 152
bond moving, 122–6
boundary conditions, 10, 28

Casimir operator, 104–5
center, group, 146, 155–7, 161
character expansion, 63, 109, 120, 156
character orthogonality, 42–3, 50, 109
charmonium, 1
chiral symmetry, 21, 26, 27, 54, 75–6, 92, 148
class function, 109, 152
condensation, 3, 149
convexity 96–100
correlation functions, 11, 36, 128
correlation length, 53, 81, 87, 144
Coulomb field, 2, 81
covariant derivative, 30, 31
creation and annihilation operators, 23, 68
critical behavior, 81, 135, 157–61
critical dimension, 118, 123–5
critical temperature, 94, 95
crossover, 139
current, 21, 76, 92

decimation, 119–26
deeply inelastic scattering, 1, 6, 90, 140
detailed balance, 130
dimensional transmutation, 86, 88–93, 151
Dirac string, 155–6

disconnected diagrams, 65, 71
doubling, 26, 107
dual lattice, 111
duality, 108–16, 126

eightfold way, 1
electrodynamics, 5, 30, 108, 118
electric field, 2, 106
Elitzur's theorem, 51, 94
ensemble, 129–30
entropy, 97
equilibrium, 127–39
Euclidian space, 14

Fadeev–Popov factor, 59
fenêtre model, 152, 155
ferromagnetism, 51–2, 117, 146
Feynman expansion, 5, 79
Feynman gauge, 91, 143
finite size effects, 128
finite temperature, 145–8
first order, 53, 99, 118, 135–8, 147, 152, 154, 156, 157, 158
fixed point, 84–7, 88, 117, 123
flux tube, 2, 54, 145
Fock space, 23
Fourier transform, 14, 15, 25, 26, 112
free energy, 18, 97–9, 139

gauge fixing, 36, 56–9, 78, 151
gauge transformation, 30, 31, 56, 58, 106
Gaussian model, 14
Gauss's law, 2, 106
generating function, 16, 70
generating state, 69–70
global symmetry, 51, 146
glueball, 53, 82, 144–5, 147
Goldstone boson, 27, 54, 56, 76, 149
grand unification, 4, 93
Green's functions, 16, 18, 68, 71, 101
ground state, 11

Hamiltonian, 8–13, 26, 58, 101–7
heat bath, 128–34
heat kernel action, 152–3
heavy ion collisions, 147

167